Galileo's Fame

GALILEO'S FAME

Science, Credibility, and Memory in the Seventeenth Century

ANNA-LUNA POST

University of Pittsburgh Press

Published by the University of Pittsburgh Press, Pittsburgh, Pa., 15260

Manufactured in the United States of America
Printed on acid-free paper
10 9 8 7 6 5 4 3 2 1

Cataloging-in-Publication data is available from the Library of Congress

ISBN 13: 978-0-8229-4859-9
ISBN 10: 0-8229-4859-1

Cover art: Cherubino Alberti, *Fame Blowing the Trumpet*, late sixteenth or early seventeenth century, gifted and dedicated to Cassiano dal Pozzo, Galileo's friend and fellow Lynx. Source: Rijksmuseum, Amsterdam.

Cover design: Melissa Dias-Mandoly

Publisher: University of Pittsburgh Press, 7500 Thomas Blvd., 4th floor, Pittsburgh, PA 15260, United States, www.upittpress.org

EU Authorized Representative: Easy Access System Europe, Mustamäe tee 50, 10621 Tallinn, Estonia, gpsr.requests@easproject.com

Contents

GALILEO'S FAME

INTRODUCTION

FAME'S VALUES

ON 13 MARCH 1610, SIR HENRY WOTTON, DIPLOMAT IN Venice, sent a short letter home to England. Wotton's juiciest piece of news concerned the publication of a curious new scholarly work, hot off the press, that had sold out within a day of its appearance. This book, Wotton held, contained such novelties that it had the potential to "overthrow all former astronomy." Yet he did not dare say whether the book's claims were true, especially as they were made with a relatively new instrument whose reliability he could not confirm. So high were the stakes, and so unpredictable the outcome, Wotton observed, that the book's author "runneth a fortune to be either exceeding famous or exceeding ridiculous."[1]

More than four hundred years later, the author of that curious little book, Galileo Galilei, firmly occupies a stellar place in our cultural firmament. He has become "exceeding famous." Nowhere is this clearer than in Florence, where the many museums hold an enormous number of portraits, statues, and busts depicting Galileo with a serious expression, a hefty beard, and an upward gaze and perhaps holding a telescope or book to signal his achievements. Visitors may stroll along the Arno to spot Galileo on the arches besides the Galleria degli Uffizi, wander along the lush, meandering Viale Galileo all the way up to the villa in Arcetri, where Galileo spent his final years, or catch his cheeky gaze in the Casa Buonarroti, where Galileo appears next to a personification of the ancient goddess of fame, Pheme or Fama, in the *studio* Michelangelo the Younger designed shortly after Galileo's condemnation in 1633 (see figs. I.1 and I.2).[2] But Galileo's presence is not limited to

Fig. I.1. Cecco Bravo, *Fama*, 1636–1637. Michelangelo Buonarotti the Younger designed the *studio* between 1633 and 1637—shortly after Galileo's conviction in 1633—and enlisted the painter Francesco Montelatici, also known as Cecco Bravo, to execute it. Source: Wikimedia Commons.

Florence or even to the other cities, such as Padua, Pisa, and Rome, that he roamed during his lifetime. His name reaches far beyond Italy, all the way to towns he never once visited, and transcends the realm of institutionalized fame, communicated through busts and street names, to appear in street art, music, comic books, and historical thrillers. Galileo has recently even been given his proper place in space: when NASA sent a spacecraft to Jupiter in 2016, on board were little LEGO figures depicting the ancient god and goddess Jupiter and Juno, accompanied by a tiny Galileo holding a globe and a telescope.[3]

The frequency and the versatility with which Galileo's image and name pop up show that the story of his achievements and trials still, more than four hundred years after his first telescopic discoveries, captivates and inspires audiences. In many cases, Galileo's name is invoked to remind people of great achievements

Fig. I.2. Cecco Bravo, *I Matematici*, 1636–1637. Galileo is portrayed amid poets, writers, geometers, discoverers, and mathematicians and holds a telescope in one hand while resting his head on the other. Source: Wikimedia Commons.

and inspire future generations to strive for similar greatness, such as when NASA and LEGO sent the Galileo figurine into space as part of an educational campaign to entice children to pursue science.[4] And then there are monuments—material or otherwise—that celebrate Galileo for his perseverance. These monuments refer implicitly or explicitly to the trial of 1633, in which the Catholic Church convicted him on the charge of "vehement suspicion of heresy." Again, Galileo's story is seen as inspirational and has been put to use in a surprisingly broad array of situations.

Still, that Galileo is so widely known and celebrated today is not self-evident. Galileo's fame did not develop organically as a result of his admittedly impressive achievements, nor because he was so apt at self-promotion. Instead, his worldwide renown is the result of a long and intense fight over his fame that involved Italian and European artists, poets, courtiers, mathematicians, astronomers, philosophers, ambassadors, diplomats, noblemen, friars, priests, and cardinals and started during his lifetime. The trial is of course the prime example of the intensity of this fight. Galileo barely escaped *damnatio memoriae* in 1633, when the formal and informal measures imposed by the Catholic Church significantly shaped the way his own contemporaries and later generations remembered him. Shortly after his death in 1642, one of his students, the mathematician Evangelista Torricelli, lamented that posthumous glory turned out to be worth very little, as Galileo had already almost been forgotten, and even now it is almost impossible to think about Galileo without also discussing the trial. However, the trial did not initiate as much as aggravate existing doubts and feelings of unease regarding Galileo and his works. Galileo, his discoveries, and his method of discovery were controversial well before 1633, as Wotton's words remind us. His 1610 discoveries of Jupiter's satellites and the moon's rugged surface went against the traditional body of knowledge, and even before that, from the earliest beginning of his career, Galileo had been involved in scholarly polemics. Strikingly, Galileo was also not the first to turn his telescope to the night sky, nor did controversies regarding his discoveries cease after the publication of his *Sidereus Nuncius*.[5] And yet, Galileo *did* achieve widespread fame—defined here as being known by people he did not know—during his lifetime. How, then, did this come about? Who shaped it, through what means and for what reasons, and what impact did it have?

This book tells the story of Galileo's fame as it developed and changed during his lifetime. It argues that his fame was not merely the direct result of his personal merit, strategic planning, and scientific achievements but the consequence of direct, conscious intervention by diverse groups of people, including many who are not immediately associated with the pursuit of knowledge. This was a period of rapid change in how knowledge was produced and disseminated, as scholars all over Europe turned their gaze to the natural world around them and compared their findings and observations about bodies, plants, and stars to what they had previously held to be true. Eager to contrast and collate their findings with other scholars, they not only sent a flurry of letters across the continent but also increasingly turned to print. Meanwhile, the marvelous nature of their findings captured the attention of audiences outside the world of academic learning. The result was not just a new appetite for learning but also the enhanced visibility of scholars as figures of public interest. While we know much about the people, places, and practices shaping the changes in knowledge production and dissemination, we do not yet fully understand the role played by local, regional, and international communities of scholars, artists, poets, and clerics in shaping the increased public recognition for new scholarly achievements and their discoverers, as well as the impact this had on the type of scholars who were held up as paragons of exemplary scholarship.

We also have not realized how intimately the concept of fame, which emerged in this period as a category connecting local reputation to later, international celebrity, was intertwined with ideas about truth and value, falsity and danger, reliability and untrustworthiness. The concept of fame, this book shows, was directly related to reputation, glory, and legal credibility but also to rumor, hearsay, and talk shared among large, mostly anonymous groups of people. Such groups were still highly suspicious and associated with rebellious intent, even as signs of public recognition increasingly came to be seen as positive. Fame thrived on this ambiguity, as the discussions it evoked spurred ever more talk, leading to more attention, visibility, and exposure. But it also meant fame was a highly ambiguous, volatile asset, especially when—as in Galileo's case—the ambiguities inherent to fame became entangled with the hopes, doubts, and anxieties surrounding new knowledge, new forms of knowledge production and communication, and new strategies for knowing at a distance. Galileo's fame became at once a focal point for people aiming to advance novel ways of knowing and a serious sign of danger for people aiming to defend established authorities. The interplay between these different views crucially shaped his life and his career. But Galileo's fame also acted as a catalyst for these wider recalibrations shaping the new culture of learning. The unprecedented public visibility of his work and his person invited engagement, controversies, and debate, and even rejections of his work, published in print or hurled from the pulpit, though meant to limit the impact of Galileo's work, further increased the public presence of scientific discoveries and practitioners.

This book follows Galileo and his supporters, opponents, and competitors from his earliest years as an aspiring young mathematician trying to find a university position, to his breakthrough in 1610, when he suddenly became one of the most famous men in

Europe, all the way to his death while exiled and semiforgotten in his villa in Arcetri. This journey takes us to universities in Padua, Bologna, and Pisa; to academies, palaces, and Inquisition rooms in Rome; and to courtly institutions and scholars' residences all over Europe, as well as to the courts of law, churches, streets, and squares of Florence and Venice. In tracing how Galileo's fame was shaped and discussed in each of these spaces, this book reveals how debates over the value of public recognition were informed by cultural, legal, and religious ways of thinking.

Early Modern Fame and Its Brokers

Galileo first became famous in the relatively short period after the publication of his *Sidereus Nuncius*. The book came out in March 1610, and in the following months news of his discoveries traveled fast throughout Europe, with early newsletters (*avvisi*) and personal letters playing an especially important part in its dissemination. These sources were particularly efficient in reaching the political and intellectual elite at courts and universities, and most European princes had become familiar with Galileo and his discoveries by the winter of 1610–1611. But Galileo's fame was not strictly confined to Europe's courts and universities. Various contemporaries reported that the streets and squares of Italian towns were rife with talk about him, he was repeatedly discussed in sermons (an early modern mass medium), and his name and telescopic discoveries were, with some regularity, used as a manner of speaking, as when the Venetian ambassador in Paris wrote that "France does not require the glasses of Galileo to see into the intentions of others."[6] His contemporaries also discussed his fame frequently, explicitly, and vigorously. At the center of their discussions, as in broader debates on fame, were questions of profit, worth, and value—in the monetary but especially in the moral sense.

For most people, fame was at once something to aspire to and to be wary of, for several reasons. First of all, fame had been widely recognized since antiquity to be a very unstable asset. Although it sometimes grew slowly and steadily and lasted for centuries, fame could also be amassed quickly and lost just as speedily. The ancient, medieval, and early modern personification of fame, the powerful goddess Fama, perfectly exemplified this duality. Her wings could

Fig. I.3. Johann Sadeler, *Allegory of Fame*, late sixteenth century. Fame transcends the earthly realm, making her way to the heavens; below on left is the moon and above on the right, the sun. Source: Rijksmuseum, Amsterdam.

Fig. I.4. Christoph Jamnitzer, *Fama*, 1610. The putto blowing bubbles—a sign of earthly, transitory fame—on the left is contrasted with the one on the right, depicting the lasting kind of glory that may be obtained through written texts. Source: Rijksmuseum, Amsterdam.

carry one all the way to heaven, but they also allowed the volatile, flighty goddess to flee rapidly and leave a person with nothing but empty air (see figs. I.3 and I.4).[7] Fame should therefore always be treated with caution, as only time would tell whether it would last and no one could expect to bend the goddess to their will.

Secondly, while fame was generally believed to be a rightful reward for extraordinary deeds, striving for it was associated with vanity and pride. This ambiguity was already present in the way ancient Roman and Greek authors thought about fame: They deemed it proper so long as it inspired one to serve the state and future generations but rejected the desire for fame as improper if one only sought to become famous for fame's sake.[8] This existing tension between seeing the desire for fame as virtuous or vain became more pronounced with the advent of Christendom. Christianity's emphasis on the need to aspire to posthumous, heavenly glory rather than earthly recognition made those striving for

Fig. I.5. Jan Collaert II, *Fama*, late sixteenth or early seventeenth century. On the goddess's banners, the means through which fame spreads—ears, tongues, and eyes—are depicted. Source: Rijksmuseum, Amsterdam.

fame during their lifetime seem especially suspect. Yet, Christian authors too recognized that the promise of earthly fame could inspire people to aspire to precisely those good deeds that would also result in heavenly glory. This ensured that earthly fame was not wholly discarded as a vain distraction from proper Christian values, although it was viewed with more suspicion than previously.[9]

A third reason for fame's uneasy position in early modern thinking related to the way it came about and was disseminated. The Latin word *fama* refers not only to the goddesses' reward, fame, but also to the process through which it developed: talk and rumor.[10] Large-scale fame was established through the accumulation of words and voices (see fig. I.5), which gave the goddess her wings and allowed fame to reach new places seemingly without effort. Again, however, fame's volatile nature marked its untrustworthiness. Such wide dissemination could only be achieved through talk shared by a large, mostly faceless group of people—hence people without clear status or authority. Although talk or chatter among larger groups increasingly came to be seen as positive over the course of the seventeenth century—a painter who was making a name for himself was said to "fa romore," that is, to be "making talk" or "making noise"—such groups also evoked suspicion, as they were just as unreliable, intemperate, and uncontrollable as the goddess herself.[11]

These three criticisms of fame—its volatility, its relation to pride, and its suspect origins—were interrelated and frequently voiced together. The problem fama's ambiguity presented was often solved, in ancient Rome as well as in early modern Europe, by distinguishing between the solid, stable, powerful, and durable fame bestowed by a small group of wise, virtuous men on the one hand (see fig. I.6) and the unstable, fickle, and volatile fame conferred by vulgar crowds on the other.[12] We indeed encounter this distinction in a letter one of Galileo's acquaintances, the Neapolitan mathematician Luca Valerio, sent him in April 1609: "*Fama* is of two kinds: the one is daughter of the common people, born by force of their stupid cries, which you rightly disprove of; the other is that which is born from a few wise men, who with their authority and natural power bend and turn the unrestrained judgment of the plebs into a sensible form: and this fame is stable and worthy of its name; the other, like an imperfect animal, arisen

Fig. I.6. Cherubino Alberti, *Fame Blowing the Trumpet*, late sixteenth or early seventeenth century, gifted and dedicated to Cassiano dal Pozzo, Galileo's friend and fellow Lynx. Source: Rijksmuseum, Amsterdam.

from the ugliness of the matter, despised by time, dies as soon as it is born."[13]

Valerio's letter illustrates two crucial points for the study of fame. The first regards its inherently ambiguous terminology. Valerio used the same word, *fama*, for both the good and the bad fame. He could have distinguished between the two through a different choice of words: *Gloria* (glory) mostly referred to the proper, long-lasting, and mostly posthumous fame that is the result of extraordinary achievements, while *vana gloria* (vainglory) referred to the fickle, empty fame only proud and vain people sought after. This difference in terminology was not always adhered to, and, as in Valerio's letter, the umbrella term *fama* often took precedence over its more specifically positive or negative forms. We are dependent on context to determine whether fama is used in a positive or a negative manner, which is especially important when investigating fame's impact, but there is a broader point here: Fame always comprised the good and the bad.

To further complicate matters, the Latin and Italian term *fama* referred to other concepts besides being known on a grand scale. We have already seen that the word also pointed to the process through which fame formed, namely talk and rumor. In addition to these, the term encompassed three further meanings as well, all of which were important in legal settings and thus point to fama's bearing on issues of credibility and truth-finding. The first of these is the legal capability necessary to act as witness or draft a will.[14] The second is very closely related to fame and often presented as its more limited, local variety: reputation. We can define this as the way someone is known and talked about, not only by those who have direct knowledge of them but also by people who obtain this information through their network.[15] The third one is public knowledge, or knowledge that is widely shared within a community.[16] These last two concepts also had their own terms in the Italian vernacular (*riputazione* and *publica fama* or *publica voce e fama*), but these were, again, frequently discarded in favor of the umbrella term *fama*.

In France and England, where the term *fama* was not used as often as in Spain and Italy, other vocabularies referred to a similar network of related and often highly ambiguous terms. In an unfinished essay on the goddess of Fame, the philosopher and statesman Francis Bacon used the word *fame* especially in its meaning of rumor, promising readers he would discuss the following points: "What are false fames; and what are true fames; and how they may be best discerned; how fames may be sown, and raised; how they may be spread, and multiplied; and how they may be checked, and laid dead."[17] Though Bacon's words reveal concerns about false rumors, they also acknowledge that "fames" could be true and that one can discern between the two (though he did not finish the part of the essay that establishes how this is done).[18] However, not distinguishing between false and true rumors was dangerous, and Bacon pointed to rumor's destabilizing potential when he wrote that "she [fame] is a terror to great cities" and that "rebels . . . and seditious fames and libels, are but brothers and sisters."[19] Elsewhere in his essays Bacon also discussed fame in another sense, which might refer to talk about one person but also to posthumous glory: "Death hath this also; that it openeth the gate to good fame, and extinguisheth envy."[20] In France, meanwhile, where the concept of reputation was intimately tied to the notion of cultural as well as financial credit, the words *los* and *renoun* were used in a broad manner, much like the Italian use of the word *fama*.[21] All these different vocabularies underline the complexities surrounding fame in this period and show how early moderns all over Europe were wrestling with the question of how much weight to assign to the talk and opinions of other people.

This, the collective nature of fame, is the second crucial point underscored by Valerio's letter. It also drives the main argument of this study. As Valerio points out, both types of fame, the valuable and the worthless, come into being through the efforts of other people, be they "a few wise men" or "the plebs." Awareness that fame developed neither on its own nor through the efforts of the person achieving fame abounded in early modern Europe. This insight has been crucial to studies that have homed in on fama's other meanings, notably reputation and talk, and how these governed community life in medieval and early modern Europe. It has also informed studies on famous early modern scientific, religious, and political figures, as historians have shown how students and scholars made Newton a genius, artists and publishers crafted "brand Luther" in Wittenberg, and Louis XIV's courtiers and advisers fabricated the persona of the Sun King.[22] What these studies do not address, however, is how this collective crafting of fame also created its highly ambiguous and unstable status, generating opportunities and risks to those in possession of it.

Focusing specifically on the brokers of Galileo's fame opens up a new perspective that highlights the limits of Galileo's own power. Mario Biagioli, in his splendid study on court culture and self-fashioning, demonstrates how Galileo carefully sought to craft his own image as the ideal Medici courtier—and that he initially had remarkable success but ultimately met his downfall due to the dynamics of patronage.[23] More recently, others have sought to trace the reception of Galileo's findings and work in different contexts. In their study of the telescope's early travels across Europe, Massimo Bucciantini, Michele Camerota, and Franco Giudice have demonstrated the importance of tending to local, regional, (proto) national, and international circumstances, while Renée Raphael's studies have alerted us to the way different communities of readers "read, used and taught" their Galileo.[24] Still others have studied the way Galileo's story came to be told after the trial and especially after his death, tracing the development of his scientific and cultural legacy to the eighteenth and nineteenth centuries.[25] This study now emphasizes how different communities, especially in Italy but also across Europe, shaped, appropriated, and opposed his fame in pursuit of their own interests from the very beginning of his career.

Galileo's fame brokers came from a wide variety of backgrounds, from academic institutions to princely courts and from literary circles to church congregations. They engaged with Galileo's fame from his earliest years to his death, thus shaping the way his contemporaries and later generations saw him. To do this, they used a wide range of media, reflecting their diverse backgrounds and the highly creative culture of science at the time: His supporters and opponents discussed Galileo not only in media traditionally associated with scholarship, such as books and letters, but also in poems and sermons, thereby reaching audiences beyond the narrow world of academic learning. Most brokers were motivated by reasons that transcended strictly scholarly considerations, showing how pragmatic, ideological, and epistemological motivations went hand in hand. The way various groups promoted and enhanced Galileo's fame among specific audiences for their own intents and purposes reveals an emerging awareness of the value of reaching different publics and of using that audience's approval to bolster Galileo's position. This recalibration of the authority of the public offers important insights into the newly developing culture of fame and the way it resembles and differs from modern celebrity.

FAME AND CELEBRITY

In recent decades, historians and sociologists have explored the nature and history of celebrity. They have debated when, precisely, modern celebrity emerged and which factors—capitalism, the development of an individual sense of self, the rise of mass media, the performative culture of theater, the transition from aristocratic court culture to the bourgeois public sphere—most directly shaped its emergence. While some scholars still pinpoint the rise of celebrity culture in the twentieth century, connecting it to the mass popularization of film and the public's collective fascination with movie stars, many now take a longer view and argue that elements of modern celebrity can already be found in the eighteenth century or earlier.[26] Thus, for Antoine Lilti the culture of celebrity emerged in the eighteenth century alongside the Enlightenment public sphere. At this point in time, a new, Romantic understanding of the individual and a novel understanding of private and public lives created not only a new category of famous persons but also an unprecedented interest in their personal lives. The rise of mass media and market capitalism facilitated and fed the public's new desire for intimate details, and, for the first time, the particular fame of celebrities transcended the boundaries of specific disciplines or groups, exerting a direct, strong, and emotional attraction on the public *as a whole*. As a result, the public became defined "by sharing the same curiosity and the same beliefs, by being interested in the same things at the same time and by being aware of this simultaneity."[27] However, Lilti argues, the public's novel, shared interest in "political debates as well as the private life of celebrities" also gave rise to its ambiguity in the eyes of elites and its conceptualization as irrational, vulgar, and problematic from the second half of the eighteenth century onward.[28]

This book instead foregrounds the fractured, fragmented nature of the public in the early modern period. In showing the long history of conflict and contestation over who deserves fame and who can bestow it, this work offers a different origin story of modern celebrity—one that may help us make sense of the way communities in our own time are fraught and divided. Galileo's fame resonated with different groups for different reasons: Florentine court poets celebrated him as a discoverer of new worlds, thereby confirming their town's cultural primacy, while members of the Accademia dei Lincei promoted him as a powerful symbol of their own dedication to observation of the natural world and a Jesuit poet in Rome praised Galileo's contributions to simultaneously highlight those of his colleague, the German Jesuit Christoph Scheiner. Meanwhile, people in Venice and Bologna initially showed themselves jealous of Galileo's success and argued that his newfound fame was undeserved and overrated. Still others rejected Galileo's fame altogether, and tensions between the various groups often ran high, resulting in vigorous accusations and attacks.

Highlighting this divisive side of early modern fame shows the contested status of celebrity has long historical roots. Debates about how celebrity culture differs from older forms of public veneration run parallel to attempts to identify the precise moment modern celebrity came into this world. As Robert van Krieken has suggested, these debates are often characterized by a tendency to distinguish between real heroes and empty celebrities and to lament the replacement of "authentic fame" with manufactured, "synthetic and empty celebrity."[29] Van Krieken suggests the dichotomy between heroes and celebrities should be rejected, not least because the fame of "true" heroes is always to some degree manufactured and disseminated by various media. This study, with its emphasis on the people who made fame, confirms that view, but it also suggests that the tendency to cast some forms of public renown as true and deserved, and others as false and empty, long precedes debates about celebrity as we know it and results from a deep-seated suspicion of the public as inherently unknowable and unreliable.

Communities were split not only over whether Galileo's fame was just and merited but also over how much public recognition was worth. Many of Galileo's supporters acknowledged that fame had the potential to attract the attention and recognition of different audiences. This was precisely what made it so enticing: Galileo's fame ensured his visibility in local as well as international settings, attracting prestige, attention, or students. His fame, or the promise thereof, made the Medici eager to employ him, made his fellow members of the Accademia dei Lincei keen to have him as a member and promote his works, and gave the Pisan *studio* a reason to pay his high salary even as he did not actually teach there on a regular basis. In each of these cases, as for most of Galileo's brokers, there was a fine awareness that latching onto and investing in Galileo's fame would yield some kind of profit—in the realm of the gift economy or the market.[30] On the other side of the spectrum, established academic and religious authorities opposed Galileo's fame with all their might. Fearful that Galileo's visibility would attract the attention of audiences that were normally beyond the reach of individual scholars and unable to properly assess his claims—such as young students and the general population of Florence—they positioned themselves as protectors of these groups. But the groups did not just need to be protected; they also needed to be protected *against*, as they, blinded by Galileo's fame and thinking they were "in the know," would easily be tempted to take to the streets, further disseminate false claims and false fames, and upset social, religious, and political order. Homing in on these concerns shows that eighteenth-century conceptualizations of the public as problematic rested on earlier views that cast specific audiences not only as especially prone to deception but also as violent, volatile, and just as difficult to control as the goddess Fama itself.

FAME IN THE SCHOLARLY WORLD

Fame was central to Galileo's career. From his earliest years to his last, it influenced his academic appointments, his social standing, and his scholarly exchanges. Galileo's lack of renown in 1588 meant his application at the Bolognese studio was rejected in favor of a better-known candidate, Antonio Magini, while his later fame proved an important asset in the context of academic appointments.[31] Thus, when in 1629 discussion arose over Galileo's high salary at the Pisan studio, where he was employed but excused from the obligation to teach, the construction was defended on the grounds of his "advanced name and fame." This, and the glory derived from his publications, would "reflect well on the *Studio*" and improve the "reputation and honor of the university itself."[32] However, this attraction Galileo's fame exercised on large groups of people was also precisely why his opponents fought it so strongly: they feared the impact it could have on the people they claimed to protect.

In discussing the impact of fame, this book ties into long-standing debates on scholarly credibility and identity. It especially seeks to deepen our understanding of scholarly communities and how clashes within these are shaped by practices and values central to institutions adjacent to scholarship, most notably the marketplace, law, and religion. Early studies on scholarly credibility highlighted the affinity between the culture of scholarship and aristocratic norms and practices, homing in on scientific academies and courts as places where knowledge was made, verified, and disseminated and situating scholarly practitioners within the noble gift economy. They also emphasized the importance of social status and of being, or at least appearing, disinterested in the outcome of research as a way of establishing credibility.[33] Subscription to gendered, aristocratic codes of behavior, Steven Shapin and Simon Schaffer have famously argued, helped English scholars to present themselves as impartial and disinterested in the outcomes of their research and thereby as more reliable.[34] Mario Biagioli's pathbreaking study on the impact of patronage further demonstrated that Galileo's ties to the Florentine court, established and forged through the exchange of gifts, boosted his status and legitimized his discoveries through their association with the Medici family.[35]

More recent works have shifted the focus away from these aristocratic spaces and instead explored scholars' ties to other spaces where credit and knowledge intersected, specifically the marketplace and courts of law. Turning his focus to Dutch scientific practitioners, Dániel Margócsy has argued that they were less concerned with appearing disinterested and more attuned to making a profit than their English counterparts. Chasing personal profit as well as epistemological gratification, they followed the rules of the market economy rather than those of the supposedly communal, disinterested Republic of Letters.[36] As a result, Margócsy shows, they preferred secrecy and competition over openness and collaboration.[37] Inger Leemans and Anne Goldgar have further explored the commercial side of scholarship by homing in on the affective aspects of the market for knowledge, highlighting the dynamic between passionate, profit-driven sellers of knowledge and interested, curious buyers. Here, interest in new knowledge was not frowned upon but actively promoted, marketed, and cultivated by stirring desire among consumers.[38] *Galileo's Fame* shows what happened when the two worlds met and long-standing aristocratic ideals of impartiality and control clashed with scholarship's growing appeal to wider publics. Under these circumstances, the very value of public recognition was as hotly contested as the novel discoveries themselves.

These debates owed much to ideas circulating in courts of law. Barbara Shapiro has turned our attention to various factors used to gauge credibility by English courts of law and by the scholarly practitioners of the Royal Society, showing that foremost among them were the expertise and experience of witnesses, as well as their reputation and rumors about them in the communities of which they were part.[39] Shapiro's findings complement those of social and legal historians focused primarily on fama's impact in late medieval and early modern Europe. They have shown that public knowledge—established through talk—could serve as the basis for starting a legal investigation. Fama could moreover serve as evidence in legal cases, as could the reputations of people involved.[40] Thus, while social status—if strictly limited to class—was perhaps less important than previously thought, social standing within one's community still mattered and was an important factor in assigning credibility. The sixteenth century in France, Andrea Frisch has argued, saw a shift in the relative weight assigned to these various factors in both the legal and the scholarly worlds. Initially, she contends, the prime indicators for reliability used in court cases were reputation and personal bonds between witnesses and judges. As the administration of justice became more centralized, emphasis shifted to expertise and experience. Frisch distinguishes a similar shift in the way authors of travel literature presented themselves before and after the advent of print. As the potential readership of books grew, the physical and social distance between authors and their audiences widened. Authors therefore stopped emphasizing commonalities between their own social background and that of their readers and instead foregrounded their experience and direct access to the events they discussed.[41]

However, the advent of print did not just entail a loss of connection between author and audience. It also brought with it the possibility of widespread name recognition, sparking interest, curiosity, and desire among readers. Published works increasingly included author portraits, and title pages prominently featured the names of their authors along with important biographical information. Some of this information, such as titles and professional affiliations, indeed conveyed an author's expertise and enhanced their authority. But including an author's place of residence and origins also helped situate authors within a specific local context, while preliminary poems, dedications, and prefaces supplied by friends, acquaintances, and patrons situated them within a social network. This need to situate famous authors within a specific local context was rooted in the value attached to reputation and social bonds within local communities; it was, in other words, tied to a need to establish familiarity, if not intimacy, at a distance. The inclusion of Galileo's Florentine origins and ties to the University of Padua linked Galileo to a specific environment and gave readers insight into his character, while also giving them a point of departure from which to check his reputation among people they knew personally. It is thus apparent that signs of local standing remained relevant even when authors reached wider, international audiences, because the markers of identity that tied authors to specific local settings helped readers with no previous knowledge of particular authors to get a sense of who they were, where they came from, and what sort of standing they enjoyed within their own communities. In other words, the mechanisms of reputation that had helped inspire trust and credibility within local communities now became important features of the emergent culture of fame, and markers of credibility used in court cases found their way into printed works. This desire for familiarity in turn preceded the need for intimacy that characterizes modern celebrity and that, as Lilti has shown, resulted in a new appetite for endless details about celebrities' private lives, habits, and diets.[42]

If details about a scholar's personal background, training, and network helped counter some of the anxieties surrounding new forms of knowledge making and transmission, the lack of information about the audience this new knowledge reached created its own problems. If the audience was largely unknown, how could their judgment be trusted? And perhaps more importantly, could those who were seen to explicitly angle for their approval be taken seriously? Scholars who obtained fame by reaching wider audiences always risked attacks on their character and credibility, which normally ran along two lines. The first common line of attack was to present fame not as the result of achievements but rather as stemming from a desire for earthly glory. Countering such claims in print was especially difficult, as this only seemed to confirm one's thirst for recognition. The corrupting effects of this desire, criticasters held, blinded Galileo to the proper rules of scholarship and spurred him to transgress them. The precise transgressions his opponents accused him of varied, but they ranged from coming up with baseless claims to get attention to stealing the ideas or inventions of others. Whatever the precise allegations against Galileo, they always presented him as out of line with the proper codes of scholarly conduct celebrated in the Republic of Letters and, as such, as unreliable. The second line of criticism went further. Fame's ability to arouse desire and spark ambition made it deeply suspect to people who sought to uphold traditional scholarship and religious stability. They argued that Galileo's fame threatened to turn the groups they sought to protect into dangerous forces of instability. Young, impressionable students would be tempted to stray from the proper academic path and be misled by Galileo's grandiose, baseless claims, while the Florentine congregation ran the risk of being infected by his pride. In this view too, Galileo's fame resulted from his misplaced and corrupting desire for earthly glory, as well as from the wider public's inability to distinguish between deserved and false fame. This meant that fame now made him not only unreliable but also dangerous.

Both fame's power and fame's daunting nature, then, resulted from its affective and collective properties. Examining how the different views on fame played out in practice in Galileo's case not only deepens our understanding of its impact on his career but also brings into focus where different ideals of scholarship meet and clash. It shows that the ideals of open, reciprocal, and disinterested scholarship were frequently violated and that even those seeking to promote such idealized learning could not entirely free themselves from earthly concerns over glory and competition. In this culture of strife, fame was used as a relatively versatile marker of credibility. It was mostly seen as a positive asset, but it could also—if circumstances made this attractive—be cast in a negative light. Conflicts over the proper interpretation of Galileo's fame, then, demonstrate the often pragmatic and deeply flexible attitude toward credibility that characterized the culture of scholarship in early modern Europe.

This book concentrates on the period during which Galileo's fame was most directly shaped: his lifetime. It does not aim to give an exhaustive account of his life and is not a biography; I urge readers looking for one to consult the excellent works of John Heilbron and Michele Camerota.[43] Rather, in the five chapters that follow I focus on several key moments that allow us to trace the development of Galileo's early reputation to his later fame and eventually to the aftermath of the 1633 trial in which his name became even more hotly contested. From the very beginning to the last stages of his career, other people played a crucial role in shaping his fama—in the sense both of his reputation and of his renown. Their efforts and motives cannot be restricted to the scholarly context alone. Practices from the law courts, as well as classical and Christian ideals, had a bearing on the formation of Galileo's reputation and fame as a scholar, and this book consequently draws on a wide range of source material, including legal documents, sermons, letters, poems, and scholarly treatises, to reflect this wide engagement with Galileo's fame.

The first chapter explores the nature and impact of reputation in legal and academic settings. Using sources pertaining to a court case, letters of recommendation, and a published report of Galileo's defense against an accusation of plagiarism, the chapter homes in on Galileo's early reputation, how this was given shape, and how it was assessed in two settings—the legal and the academic. The chapter reveals that Galileo's reputation in this period was largely positive but never wholly uncontroversial: in both the legal and the academic contexts rumors circulated that cast doubt over Galileo's honesty. This was particularly problematic, as Galileo's reputation—in both contexts—was firmly linked to his credibility. The chapter further demonstrates that these doubts could be negated only through the personal testimonies of patrons, friends, and other members of the communities to which he belonged.

Chapters 2, 3 and 4 each address fundamental mechanisms underpinning the development of fame: competition, appropriation, and opposition. Chapter 2 turns to the period immediately following the publication of Galileo's *Sidereus Nuncius*. In 1610, Galileo's telescopic discoveries caught the attention of new audiences in all of Europe, but those audiences initially struggled to determine their value. In this situation, the growing desire for more information about Galileo quickly became entangled with a strong sense of competition that played out on different levels. Those unfamiliar with Galileo reached out to people they knew who did have access to him or at least knew him by reputation, and, this way, the local reputation Galileo possessed before reaching more widespread fame played a crucial role in the way new audiences responded to him. Meanwhile, when those familiar with Galileo communicated their opinions to others, they focused on one element in particular—Galileo's Florentine background—to comment on his character and his desire for glory, using this to make further claims about his credibility. Their comments themselves, the chapter shows, were also shaped by competition over glory.

The third chapter focuses on three different types of poets who celebrated Galileo and his discoveries, to show that admiration and appropriation went hand in hand. All three groups—a group with ties to the Florentine court, a Jesuit philosopher in Rome, and Galileo's fellow members of the Accademia dei Lincei—were very strategic in seeking publicity for Galileo and his work, doing so only if they thought they stood to gain enough from it. The extent of their praise depended on the poets' proximity to Galileo, their personal circumstances and interests, and the success he seemed to enjoy at the time he was writing. By highlighting that the poets frequently adapted or even appropriated the story of Galileo's fame to fit their own purposes, rather than simply amplifying the narrative Galileo himself had created, the chapter also points to an important risk of fame: its recipient's loss of control. Finally, the chapter shows that poets praising Galileo drew a firm connection between his fame and his pursuit of novelties, which they praised as heroic, adventurous, and worthy of praise, thereby providing an important legitimation for new scholarship.

This was not the case for the main protagonists of chapter 4, which focuses on the opposition Galileo's fame evoked. The academic and religious opponents who attacked Galileo in scholarly treatises and sermons instead drew a firm connection between Galileo's perceived desire for glory, his (sinful) vanity and pride, and his unreliability. Besides highlighting similarities in the arguments used, the comparative approach central to this chapter also points to parallel motivations among Galileo's academic and religious opponents. Both groups of criticasters perceived Galileo as a threat to the communities they felt responsible for and feared might become corrupted if he went unopposed. In highlighting these concerns, the chapter probes the relation between rumor and fame and shows that Galileo's opponents perceived his growing following as a menace to social, political, and religious order. As a result, they launched attacks on him using the best means available to them: academic treatises for the scholarly opponents and the pulpit for his religious criticasters. Finally, the chapter also reveals a crucial difference in the dynamic of fame as it played out across these two contexts. Galileo was able to reply in kind to his academic attackers and as a consequence did not suffer much from them, but he had very limited means to respond to the charges launched from the pulpit.

At this point, there is a jump in the narrative, as the final chapter probes the effect of the trial of 1633 on Galileo's fame, posthumous glory, and memory. I argue that the sanctions taken against Galileo can best be understood as attempts to redirect Galileo's fame *and* the way he would be remembered: not as a brilliant scholar but as a testament to the Catholic Church's authority. The chapter also reveals that the community of scholars targeted by the Catholic Church resisted its efforts and created paper and material monuments to preserve Galileo's good name for future generations. The chapter thus highlights the strong dedication to Galileo's good name among a small group of supporters in light of the pressure exerted by a powerful institution. My examination of the monuments they created demonstrates this was by no means an easy feat, but Galileo's supporters did employ at least five different strategies to praise him without directly going against the sentence of 1633. This way, they laid the groundwork for the way Galileo is remembered to this day—a theme that is taken up in the epilogue.

This study of books and talk, portraits and poems, sermons and witness testimonies, and above all the communities that made and contested Galileo's fame illuminates the highly ambiguous nature of fame in the early modern period, showing how it elicited not only admiration but also envy and criticism. In a period when the value of public recognition itself was still being negotiated, fame was at once something to covet and something to fear, and it was in the long struggle between those who supported him and those who sought, with all their means, to oppose him that Galileo's fame took its precise shape.

1

Before Fame

Galileo was born in Pisa in February 1564 to Giulia Ammannati and Vincenzo Galilei, a lute player and musical theorist. Though Vincenzo was of noble status, his income from making and teaching music was too limited and unstable to make the family wealthy, and, hoping that his son would fare better in this regard, he enrolled sixteen-year-old Galileo at the university in Pisa to study medicine. The youth did not have a knack for the subject and instead decided, around 1583, to become a mathematician. After leaving the university without a degree in 1585, Galileo spent several years honing his literary, artistic, and mathematical skills among Florence's noble elite, enjoying the rich cultural life Florence and Pisa had to offer while struggling to find a university teaching post. He made ends meet by teaching privately until he found a post in Pisa in 1589, after which, in 1592, he moved to Padua. Though he occasionally corresponded with other mathematicians abroad, Galileo's formative years were especially shaped by the local environments of the towns where he resided in the first four and a half decades of his life: Pisa, Padua, and Venice.[1] If not for his work with the telescope and subsequent celestial discoveries, which catapulted him to international fame in 1610, Galileo might simply be known to history as yet another provincial mathematics professor and closet Copernican with whom Johannes Kepler had exchanged some letters.

Yet, before gaining international renown, Galileo possessed a different type of *fama*, which, although related to fame, was not entirely the same. It entailed his personal reputation, or what people who knew him said and knew about him more or less directly. This reputation, established through talk and letters, was far more limited and local in scope than the widespread renown we call fame. Nonetheless, it was a vital prerequisite to fully participating in early modern societies. Galileo's reputation affected his social relations, his job prospects, his finances, and his credibility, both as a scholar and as a witness. Historians have long shown that the reputation of witnesses played a crucial role in court cases, where it was used to weigh the credibility of testimonies. It is, however, much less clear how much weight was assigned to reputation within the world of scholarship; historians have especially debated the relative importance of reputation, social standing, and status vis-à-vis other indicators of credibility, in particular expertise and experience.[2] Galileo's case presents us with an opportunity to study the reputation and credibility of one person, over a relatively short period of time, in the contexts of both law and scholarship. To that end, this chapter explores Galileo's early reputation through three separate episodes in the first half of his life. While this approach leaves out some of Galileo's early achievements, the three cases here offer us valuable insights into the nature, impact, and limits of Galileo's reputation within the contexts of the law court, the academic job market, and a scholarly polemic and plagiarism conflict.[3]

In all three settings, Galileo's reputation was formed through the efforts of others, who—orally and in writing—testified to his character, proficiency, and reliability. Yet, the impact of their efforts differed from context to context. Within the legal setting, judges used his reputation to assess the credibility of Galileo's testimony, while also taking additional factors, such as disinterestedness, social status, and his extensive knowledge as eyewitness, into account. Galileo's reputation as a proficient and honorable mathematician was similarly shaped by authoritative acquaintances in academic settings. Here, however, a good reputation did not always suffice, and to further his career he needed other forms of fama. Galileo especially encountered the limits of his good reputation as he was rejected for an academic position on the grounds of his lack of renown, highlighting the important role fame already played in universities' hiring practices in the early modern period. Finally, as Galileo's renown slowly started to grow after he obtained a position in Padua and published a first work, he also experienced the perils of this enhanced visibility, and academic and legal practices converged to defend his good name and the modest fame he had achieved.

REPUTATION IN THE COURTROOM

Galileo's early reputation, court records show, was not wholly uncontroversial—and, consequently, neither was his credibility. When Galileo was about twenty-six years old, he was called before the Arte dei Giudicai e Notai, the guild of the judges, lawyers, and notaries.[4] Galileo was neither plaintiff nor defendant but appeared as one of the main witnesses in a case regarding the thorniest of issues: an inheritance. Giambattista Ricasoli, one of Galileo's best friends, had changed his will shortly before passing away in January 1590, leaving his estate to a distant cousin rather than to his sister Maddalena. As he had suffered from manic episodes and paranoid tendencies, his sister and her husband, Jacopo Quaratesi, argued that Ricasoli had been vulnerable to manipulation. They therefore sought to have the latest version of the will nullified.[5] The ensuing investigation opens up a window through which to study Galileo's early reputation and how this related to his credibility.

The court, tasked with the decision to either uphold or nullify Ricasoli's will, used different strategies to determine whether Ricasoli had been of sound mind. One of them was to rely on the accounts of eyewitness testimonies, among which was Galileo's. He had often accompanied Ricasoli when the latter fled his native Florence to surrounding towns and cities, and the court wished to know how Ricasoli had behaved in his presence. But Galileo's interrogators were careful not to assign too much weight to eyewitness accounts, using them mostly in conjunction with another form of knowledge: common knowledge, or *publica fama*. Finally, the interrogators also carefully considered the credibility of each individual eyewitness, for which they used several indicators—one of which was reputation. At the heart of the case, then, was the question of how much weight to assign to different forms of fama.

The investigation started with the claim, put forward by Ricasoli's sister and her husband, that it was publica fama, or publicly known and said, that Ricasoli was not of sound mind at the time he changed his will. Publica fama constituted the sort of public or common knowledge shared within communities: the thing everyone has heard and talks about, knowing it to be true. This sort of knowledge had two important functions within the early modern legal system: It could serve as the basis to open an investigation and as evidence within such procedures. But the probative value of such knowledge was never self-evident, as the notion of publica fama was closely related to that of fama in the sense of talk or rumor, which was often viewed with suspicion because of its uncontrollable nature. Awareness that publica fama could be manipulated and that conflicting accounts could exist within one community further complicated the status of public knowledge in court cases.[6] To determine whether to nullify or uphold the will, the interrogators in the Ricasoli case would have to determine whether Ricasoli's mental instability was indeed publicly known (they consistently used the phrase *publica voce et fama*, or "public voice and talk") and whether it was corroborated by further evidence. To this end, they asked all witnesses to relay what they had heard said about Ricasoli and to comment on the nature of publica fama itself.

Galileo's testimony corroborated the claims by Maddalena and her husband and furthers our understanding of publica fama in court cases, as Galileo's answers point to the various strategies that could be used to distinguish rumor from truth. One of them was to rely on numbers. As Galileo put it, when asked to define publica fama, "It's publica voce et fama when everyone, or the major part, says the same thing, and for the rest refer to the laws."[7] Still, quantity was no guarantee for truthfulness; it also posed risks. Frequent iterations of a piece of knowledge offered a sense of security, but its dissemination among a large group could also obscure the original source. This meant it became difficult to verify whether such publica fama was based on more than just talk.[8] One way to overcome this problem was to establish whether the common knowledge was new or old. As Galileo told his interrogators, "It is almost impossible that a publica voce et fama, continued for a long time, is false."[9] He also told his questioners that because Ricasoli's reputation as being of unsound mind stemmed from November 1588 and had not thereafter been contradicted, he saw no reason to assume that the common knowledge in this case diverged from the truth.[10]

But the most important way of establishing the veracity of publica fama consisted of tracing it back to its origins. The interrogators therefore asked Galileo how the publica fama about Ricasoli had originated, "which actions have been observed and recounted by persons worthy of trust, which may make up the aforementioned publica voce et fama, and who have been those persons worthy of trust, and were they gentlemen or of what quality of persons, and are they above any objection."[11] Galileo answered as follows: "The actions that led Giambattista [Ricasoli] to become reputed as oppressed by melancholic humors have been observed and recounted by multiple people worthy of trust, and in particular by Mess. Lorenzo Giacomini, Mess. Francesco Guadagni, Mess. Iacopo Quaratesi, Mess. Neri Ricasoli and Giovanni his brother, Mess. Giovambattista Strozzi, Mess. Bernardo de' Bardi, Mess. Giulio da Barga physician, Mess. Agnolo Bonelli physician, a doctor who treated him in Genoa, a *Teatino* from Genoa, and by others whom he [Galileo] does not recall at this moment."[12]

The relation between publica fama and eyewitness observation Galileo drew here shows that these two types of evidence were not necessarily sharply distinct or differently valued types of knowledge. In ideal cases, publica fama was the widely shared oral reflection of actions that had been directly observed by one or more eyewitnesses. In such cases, its reliability rested both on the extent to which it was shared and on how precisely its initial origin could be pinpointed. The identification of a specific claim's precise origin gave the court a further way to assess that claim's credibility: It could now also take into account the credibility of the initial observer(s).

In Galileo's case, the court first probed the extent and depth of his knowledge. Over several days they asked him more than two hundred questions, from the very general to the highly detailed, all to determine whether Ricasoli's behavior had been out of the ordinary. They asked Galileo to comment on his friend's use of language, his habits of worship, the degree (appropriate or not) of modesty when approaching his peers, how he wore his beard, and whether he lavishly spent money or was on the whole a frugal man.[13] Galileo's answers revealed his extensive knowledge about Ricasoli's habits and changes therein. He provided the court with insights into their most pointed questions and was confident in his answers, as when he told them he did not find it strange that his friend slept only in a nightshirt, without a head covering, during the hot months of July and August. He also freely admitted gaps in his knowledge, as when his questioners wanted to know whether Ricasoli thought Naples more beautiful than Rome, a rare question to which Galileo did not know the answer.[14]

On the whole, then, Galileo's testimony was nuanced, detailed, and convincing. This did not mean that the court unconditionally accepted his account. Besides probing the depth and extent of knowledge of each individual witness, the interrogators also sought to establish whether the witnesses themselves were *degno di fede*. Not every witness was to be believed, or even heard. Only those who possessed fama in its meaning of legal status or competence, in some cases called *fama legalis*, had access to legal capacities, such as the ability to testify. Most individuals possessed this status as a matter of course (though it was denied to some, such as bastards), but it could be lost. This could happen after, for instance, being convicted of certain crimes.[15] The phrase had a second, more general meaning as well. Credibility was not evenly distributed among those who possessed fama legalis: some were deemed more reliable than others.

We have already seen that the court, wanting to know where the publica fama about Ricasoli originated, asked Galileo whether the persons who had originally observed his actions were gentlemen. This emphasis on social status confirms earlier studies, which, especially for the academic setting, have argued for the importance of nobility as a marker of credibility. But nobility was not the only marker used, nor was it decisive, as Galileo's answer to the court shows. He echoed their preoccupation with social status but added a second indicator of credibility as well: He included, when relevant, the original eyewitnesses' professions. Some of them were physicians, and they of all people should recognize what constituted proper versus confused behavior. And, asked if one should "grant more credence to four gentlemen, than to servants, coachmen, hosts or apprentices," his confirmation came with an important caveat: Gentlemen should only be believed more than the others "if the facts can be known to all the people mentioned in the question."[16]

Noble status alone was not sufficient, which also becomes

apparent from the court's assessment of Galileo's credibility. To determine how much weight to attach to his testimony, the court used information about his family background and education alongside rumors about his character. This approach did not work in Galileo's favor. Various lists of witnesses drawn up by the court, which briefly summarized each witness's most important features and possible objections to their testimony, demonstrate Galileo's dubious credentials in this period.[17] One list referred to Galileo as "defrocked, son of a music teacher," while another one characterizes him as "Galileo. Was a monk of Vallombrosa, son of a teacher for lute playing," and on a third list he dismissively appears as "Galileo Galilei, son of a lute player, poor and defrocked, etc."[18] The court drew up these lists by asking witnesses what they knew about each other. This way, talk from within a community—which in Galileo's case was not wholly positive—found its way into the courtroom, where it was used to weigh each witness's testimony. The information about Galileo may have reached the court through Giovanni Ricasoli, the defense party who was the main beneficiary of Ricasoli's latest will, or his lawyer, who seems to have tried to undermine Galileo's testimony. Such a motivation is suggested by the marginalia in the witness reports, which repeat the characterization *sfratato* several times and in different forms: "Galileo defrocked," "the defrocked," and sometimes in a longer sentence: "or he was an unfrocked friar without discretion."[19] These slights likely referred to the period Galileo spent in the monastery of the Vallombrosan order, where Galileo was sent to study when he was about eleven years old and which he left after about four years.[20] His leaving the order, while most likely a practical decision, could be used against him as it implied a lack of perseverance. It also suggested he was not above breaking his vow—hardly a commendable quality for a witness. The slight is repeated in the margins alongside a passage where Galileo admits he momentarily left Giambattista Ricasoli at a time when he needed him, further casting doubt on his reliability as witness and a friend.[21]

Galileo's poor financial status lent credibility to the charge that he might be bribed. What was worse, Giovanni Ricasoli even accused Galileo of stealing from Giambattista Ricasoli on a trip to Milan, claiming that it was "true, publicly and well known" that Giambattista then sent Galileo back to Florence. The annotations on the interrogation records show that the defense party also accused Galileo of being bribed by Giambattista's sister Maddalena and her husband Jacopo, possibly to help place one of his sisters in a monastery. One annotation mentions that "they have offered you 150 florins for the sister, to do this etc., wrongly!," while another states, "These things no one says, if not with the goal of making his sister a nun."[22] These accusations, which Galileo of course denied, mounted major assaults on Galileo's credibility as a witness: They portrayed him as a dishonest thief, one who furthermore had a personal, financial interest in the outcome of the trial.

Yet, the interrogators, taking into account that discrediting the other party's witnesses was a fairly standard courtroom practice lawyers employed, did not simply accept the claims. As they had with the publica fama about Giambattista Ricasoli, they again set out to investigate whether Giovanni Ricasoli's claims were corroborated by evidence or by other witnesses. They were especially cautious because Giovanni, being the main beneficiary of Giambattista's revised will, certainly had a financial stake in the outcome of the trial. They therefore asked another witness, the nobleman Lorenzo di Jacopo Giacomini Tebalducci, to reflect on the accusations against Galileo. Tebalducci testified that he "did not believe in any way" that Galileo had supposedly stolen fifty florins from Giambattista Ricasoli. The interrogators also inquired after the other bits of information they had gathered on Galileo, asking Tebalducci whether it was true that Galileo's father headed up a music school, was poor, and had many children. Tebalducci confirmed the first claim but said that he knew nothing about Vincenzo Galilei's income or children, thereby allaying some of the suspicions. Most importantly, he firmly dismissed as slander (*calunnia*) the idea that Galileo had hoped to secure a place in the monastery for his sister and might therefore have testified as a favor to Maddalena and Jacopo.[23]

The slights Giovanni Ricasoli had raised against Galileo's character were thus abated by another witness, showing the continued relevance of personal ties as a way of establishing credibility. Talk within a community found its way into the courtroom and impacted the reliability courts assigned to its witnesses; even eyewitness testimonies were weighed through this process. The thoroughness with which the court checked Galileo's credibility resembles its way of working at large. Cautiously corroborating hearsay and eyewitness testimonies, the court ultimately decided the conflict over Ricasoli's estate in favor of Maddalena and her husband in September 1591.[24] The decision was the result of an intricate system of knowledge-finding in which various types of knowledge were used together; publica fama and eyewitness reports were not seen as sharply distinct types of evidence, nor was there a strict hierarchy of them. Instead, they interlocked and were used in tandem, as were the various indicators the court relied on to probe the credibility of individual witnesses.

The Limits of Reputation

In the same period the Florentine community of witnesses discussed Galileo's reputation in the courtroom, another set of witnesses shaped his academic reputation. Between 1588 and 1592 Galileo applied for university positions at Bologna, Florence, Pisa, and Padua.[25] Much like in the legal setting, in the academic context Galileo's reputation as a proficient mathematician was shaped through the efforts of people who acted as brokers on his behalf. Focusing especially on Galileo's application for the vacant chair in mathematics at the Bolognese *studio* in the early months of 1588, this section explores the strategies Galileo and his supporters had at their disposal to convince universities to hire him—and why they were not effective in this particular situation, even if they might have been in different circumstances.

Galileo's application to the Bolognese studio consisted of two recommendation letters, each very different in tone, and a set of mathematical theorems. The first of the letters drew heavily on the language of friendship, favors, and gifts. It was written by Cardinal Enrico Caetani, whom Galileo may have met, possibly through Christopher Clavius, during his trip to Rome at the end of 1587.[26] Caetani was not an expert mathematician and seems to have only been very superficially acquainted with Galileo. His letter stated that he "has great approval of his sufficiency" but did not otherwise go into detail about Galileo's mathematical skill or previous expertise. Instead, his recommendation derived its weight from his status. Caetani had recently been appointed camerlengo by Pope Sixtus V but served, between 1586 and 1587, as papal legate to Bologna. Bologna's university was the oldest and, until the late sixteenth century, when Padua took the lead, the most renowned university in Italy. The responsibility for the recruitment and appointment of professors lay with the Assunti di Studio, a group of four senators, but their decisions were subject to formal approval from the senate and the papal legate, who acted as patron to the university. The latter was frequently enlisted to help in the fierce competition to attract particularly distinguished professors, and Caetani wrote from a position of authority when he recommended Galileo. He took care to emphasize that he would deem it a personal favor if the studio were to appoint Galileo: "If my recommendation of him be of use to him in the competition with others to approach Your kindness, I will feel particularly obliged to you."[27] The form and tone of Caetani's recommendation fit his station, but stressing Galileo's ties to people with power did not guarantee success; the relation between the Bolognese local authorities and those in Rome was often tense, and the senate was wary of Roman attempts to increase its influence over the university and, by extension, the city.[28]

The second statement in support of Galileo took a wholly different approach. This one, by a Signor Artani, was most likely delivered (either orally or in the form of a letter) to the Bolognese senator Giovanni dall'Armi and passed on to the rest of the senate in late 1587. The testimony suggests a closer relation to Galileo and included biographical details about him, such as his noble status and his age, and it carefully stressed Galileo's experience and expertise.[29] Artani pointed out that he had studied with Ostilio Ricci, was educated in "all the mathematical sciences," and had, moreover, "great judgment in this and many other subjects that he has studied, especially in the humanities and philosophy and other good qualities." To further increase Galileo's chances of getting the job, Artani emphasized that Galileo already had experience teaching mathematics, telling the senate that Galileo "was appointed to the public lectureship of Mathematics in Siena; and he has also practiced privately, and has read to many gentlemen both in Florence and in Siena."[30] Artani may have taken some liberties with the truth to increase Galileo's chances, as no evidence has come down to us that Galileo had delivered public lectures before 1588, when he delivered a series of lectures on the measurements of Dante's hell for the Accademia Fiorentina.[31] Artani also added a few years to Galileo's age—he was twenty-three, not twenty-six, as Artani claimed. For a starting university professor at the time, twenty-three was not exceptionally young, and universities often "reserved" a limited number of teaching positions for recently graduated students.[32] However, the chair Galileo applied for was not such a position. In this case the university sought someone with a more established record and profile, and Artani likely thought that the years added to Galileo's age made him appear a slightly more convincing candidate.

Galileo's recommendation letters thus took up complementary strategies, but together they communicated to the Bolognese senate that Galileo was not only well connected but also experienced and proficient. The final aspect of Galileo's application consisted of a set of theorems, in the Archimedean mechanical tradition, that he developed between 1586 and 1587, with suggestions and feedback from two mentor figures: the Marquis Guidobaldo del Monte and Christopher Clavius, the head of the Jesuit mathematicians at the Collegio Romano. In the theorems, Galileo sought to demonstrate the precise locations of the centers of gravity of specific, solid objects, thus contributing to the practical application of mechanical knowledge. Both Clavius and Del Monte fought hard to raise the status of mathematics as a discipline, and each appreciated Galileo's work while also suggesting, when Galileo first shared them, that the theorems needed further improving. After reworking them to his own satisfaction (Clavius was still unconvinced), Galileo found four Florentine gentlemen willing to testify that he was the author of these theorems, thereby converting them into a symbol of his mathematical proficiency and his social connections.[33] Two weeks later Giuseppe Moletti, public lecturer of mathematics at the Studio di Padova, added a fifth formal endorsement. Moletti confirmed that Galileo had discovered these theorems and additionally commented on their quality: "I, Giuseppe Moletti, public lecturer of Mathematics at the Studio di Padova, say that I have read the present lemma and theorem, which have seemed correct to me, and I judge their author to be a good and practiced Geometer."[34]

In normal circumstances, Galileo's approach might have worked. This time, however, it did not. In April 1588, the senate finally responded to Caetani's letter, sent in the second week of February. The senators confirmed that they had been looking to appoint a new professor but that they had not discussed the search for a suitable candidate for some time and had halted the search temporarily. The senate ended the letter by assuring Cardinal Caetani that they "would keep [Caetani's] recommendation and [Galileo's] good qualities in mind." It was a polite refusal of Caetani's recommendation and Galileo's application; shortly after rejecting Galileo, the university hired the Paduan Antonio Magini to fill the position.[35]

Magini was a better candidate than Galileo: He was nine years older than Galileo and had been educated at the Bolognese studio, giving him a head start over his rival for the job. He also possessed a much broader range of expertise. The previously separate chairs of astrology, astronomy, and mathematics at the Bolognese studio had toward the end of the sixteenth century been replaced by a single chair for mathematics.[36] It is ironic given his later achievements, but in 1588 Galileo was not known for his interest in astronomy. In contrast, Magini had published a work on ephemerides in 1582, indicating his ability to instruct students in all mathematical disciplines. Magini's publication record proved crucial in another sense as well. In their rejection of Galileo, the senators explicitly told Caetani that they had enough "normal" professors and were looking specifically for someone of "nome et fama."[37] The word *fama* here referred not to reputation in local settings but to the wider renown that meant someone was well known beyond their immediate personal and professional milieu. This desideratum was a matter of quality and quantity alike: a professor with sufficiently positive and broad name recognition among a wide audience would entice new students to attend the studio, which in turn drew much of its status from its student numbers.[38] When it came to fame, Magini's publications spoke strongly in his favor. Galileo had not yet published anything, but Magini's Italian and Latin publications had the potential to attract local as well as international students to the studio.[39]

The failed application shows the limits of Galileo's reputation and of personal brokerage. While precisely what was needed to establish fama in the sense of reputation, personal recommendations generally fell short when fama in the sense of renown was concerned. Exceptions exist, of course; published recommendation letters and publicly shared recommendations in any other form (such as the laudatory poems discussed in chapter 3 of this book) could both serve as effective means to gain renown, reaching larger audiences than a personal recommendation directed to one person or a specific institution. Normally, however, their audience was too limited to make sufficient impact in this regard.

Galileo did, around this time, try to increase recognition for his work. One step he took was circulating it more widely, specifically among other mathematicians. These efforts earned him the praise of one of his earliest patrons, Guidobaldo del Monte, who in late 1588 reached out to say that he was happy to hear that Galileo planned on circulating his work on the centers of gravity and believed Galileo would certainly acquire honor in doing so.[40] Galileo also tried to expand his network, especially through a trip to Rome at the end of 1587. This led to his connection with Christopher Clavius, arguably this period's most distinguished mathematician—he was known as the Euclid of his time—and to his first foreign endorsement.[41] The famous geographer Abraham Ortelius, then residing in Rome, encountered Galileo's work (possibly through Clavius) and forwarded a sample to Michel Coignet. Coignet, a renowned mathematician working in Antwerp, then expressed his approval of Galileo's theorems in a letter written in March 1588.[42] Such international connections were an important asset for building renown, as Coignet could recommend Galileo—much like Ortelius had—to others in his own network whom Galileo had not yet met. This outward spreading of knowledge about a person created the asymmetrical relationships typical of renown, in which individuals are recognized by people they do not know personally.

In his search for other academic positions, Galileo continued to rely on the personal recommendations of authoritative figures. The brothers Guidobaldo and Francesco Maria del Monte, members of an important noble Tuscan family, were particularly important as he applied for jobs in Florence and Pisa between 1588 and 1589. The university at Pisa was much smaller than the Bolognese or Paduan ones and attracted fewer international students, making it less concerned with hiring famous professors and more prone to hire locally (in contrast to Padua). The Medici grand duke, moreover, exerted direct influence over appointments via the studio's overseer. The Del Monte brothers were quick to congratulate Ferdinando I Medici on becoming grand duke in 1587, and when Francesco Maria del Monte was elected cardinal in 1588, the circumstances were particularly fortuitous for Galileo; his patrons were now in an excellent position to recommend him to either the grand duke or directly to the studio. He was appointed to the Pisan studio in the summer of 1589.[43]

For his Paduan application three years later, Galileo sought out the help of the local nobleman Gian Vincenzo Pinelli. Pinelli was one of the city's leading intellectuals and had surrounded himself with a considerable network of scholars. Among his acquaintances were various administrators of the university in the city, which came in handy as Pinelli lobbied for Galileo's appointment to the studio in September 1592 by tracking them down and listening attentively to accounts of the ailments they were suffering before slipping in a kind word about Galileo.[44] The strategy worked, and Galileo was awarded the post at Padua, where he would remain until 1610. The Paduan studio was about half the size of Bologna's, but it steadily gained renown over the second half of the sixteenth century. Venice's strong support of the Paduan studio meant that the position paid much better than Galileo's previous post in Pisa, but this would be only one of the reasons Galileo later called the years he spent in the Venetian republic the happiest of his life. These were also the years Galileo spent in the company of Marina Gamba, the woman who is deplorably absent from his correspondence but with whom he had two daughters, Virginia and Livia, born in 1600 and 1601, and a son, Vincenzo, born in 1606.[45]

Galileo also significantly expanded his scope as a mathematician during the years he worked in Padua, laying the foundations for some of his later work. The city's proximity to Venice proved important when the Arsenal's *proto* consulted him on practical problems related to Venice's long-standing attempts to improve the performance of its large ships in military encounters with the Ottoman fleet. Though Galileo did not provide useful solutions to perfect the placement of rowers and oars, his investigations on the subject did lay the groundwork for his science of the strength of materials, which would make up a large chunk of his last published work, *Discorsi e dimostrazioni matematiche intorno a due nuove scienze*, printed in Leiden in 1638.[46] Galileo also took his first serious steps in astronomy in this period. Following the appearance of a new star—what is now called a supernova—in the heavens in October 1604, the university asked him to provide a series of lectures on the subject. The star's sudden appearance led to anxiety on the part of those who interpreted it as a bad omen, but the supernova also called into question the Aristotelian—and Christian—idea that the heavens were perfect and immutable, causing unrest among astronomers and philosophers both. This made the topic at once highly fascinating and highly controversial, and the three lectures Galileo delivered from the aula of the Palazzo Bò are sometimes said to have drawn more than a thousand spectators. In his own lecture notes he referred only to "numerous youths who have flocked here" to hear him speak, but there is no denying that the supernova caught the attention and interest of many.[47]

In his lectures, Galileo defended the idea that the star was not so close to the earth as to belong to the sublunar realm but must be located *above* the moon. Not long after, various pamphlets appeared that defended Aristotelian cosmology and criticized Galileo's mathematical approach. One of the pamphlets was by Baldassare Capra. His 1607 conflict with Galileo is the focus of the next section, but already in 1605 Capra felt that Galileo had not properly credited him and his teacher Simon Mayr with first observing the supernova in Padua. Another may have been written by Cesare Cremonini, a professor of philosophy at the studio with whom Galileo had a friendly but complicated relation. Galileo likely ignored Capra's attack, but he answered Cremonini's criticism with a pseudonymous pamphlet of his own, published under the name Cecco de Ronchitti ("Blind man with the coughs" or "Blind man of the blind alleys"). Written in the form of a dialogue between two Paduan peasants, the pamphlet poked fun at Aristotelian philosophy in the strongest terms, and it did so in the vernacular—making the work accessible to people who did not read Latin. The exchange of pseudonymous pamphlets by two professors of the Paduan university attests to the particular culture of liberty in the Venetian republic in the early seventeenth century: Though the Venetian authorities maintained and supported an unusual degree of freedom vis-à-vis Rome, it was nonetheless wise to conceal one's true opinions behind a mask.[48]

Fig. 1.1. Domenico Robusti (called Tintoretto), portrait of "Gallileus Gallileus Mathus:," ca. 1597. © National Maritime Museum, Greenwich, London.

The Paduan studio was much more internationally oriented than the Tuscan institutes of higher learning, but Galileo seemed more interested in cultivating close ties with the local men of learning than with other mathematicians across Europe. Galileo happily taught the young European noblemen coming to his house for instruction, but when his work piqued the interest of international scholars such as Tycho Brahe and Johannes Kepler in the late 1590s, he did not take up their invitations for intellectual exchange.[49] Instead, he maintained especially good relationships with the circle around Pinelli and struck up close friendships with Gianfrancesco Sagredo, a member of Venice's Great Council, and Paolo Sarpi, a Servite friar, mathematician, and freethinker. Sarpi would come to play an especially important role in Galileo's career: He was the first to alert him to news about the new instrument that would later become known as the telescope.[50]

Galileo's growing standing within the Venetian republic's cultural elite is also suggested by a portrait, made around 1605 or 1606, when Galileo was in his early forties. If the dating is correct, it was made around the time Galileo published his pseudonymous pamphlet on the supernova and his manual on the geometric compass. The painting depicts Galileo looking directly at the viewer with a scrutinizing and slightly combative expression, unsmiling and stern (see fig 1.1). It is usually ascribed to the Venetian painter Domenico Robusti, called Tintoretto (like his better-known father Jacopo), who from the late sixteenth century onward painted portraits of many Venetian noblemen as well as foreign princes. An inscription in the upper-left corner of the portrait identifies the sitter as Galileo Galilei, mathematician, "Gallileus Gallileus Mathus:."[51]

The publications, the portrait, and the international appreciation from Kepler and Brahe all indicate that Galileo's visibility and standing slowly but steadily grew during the years he spent in Padua and Venice. This new phase in the development of Galileo's fama, however, was not without its risks.

AN ACADEMIC TRIAL

For Galileo, the worlds of the law court and the university truly converged in 1607. That year a former student plagiarized one of Galileo's works, leading Galileo to lament the usurpation of his "honor, reputation, and deserved glory."[52] But rather than taking up the arms of the scholar, namely pen and paper, Galileo responded by orchestrating a full-blown trial, overseen by officials from the Paduan university, in which he asserted his mathematical superiority over the student foolish enough to attack him. To state his case, he relied on the help and support of authoritative acquaintances, mostly from Tuscany and Venice, whose testimonies were

DIFESA
DI GALILEO GALILEI
NOBILE FIORENTINO,
Lettore delle Matematiche nello Studio di Padoua,

Contro alle Calunnie & impoſture
DI BALDESSAR CAPRA
MILANESE,

Vſategli sì nella Conſiderazione Aſtronomica ſopra la nuoua Stella del M DC IIII. come (& aſſai più) nel publicare nuouamente come ſua inuenzione la fabrica, & gli vſi del Compaſſo Geometrico, & Militare, ſotto il titolo di

Vſus & fabrica Circini cuiuſdam proportionis, &c.
CVM PRIVILEGIO.

IN VENETIA, M DC VII.

Preſſo Tomaſo Baglioni.

Al S. Niccolò Mannozzi l'aut.re

Fig. 1.2. The title page of Galileo's *Difesa*, with Galileo's inscription (*bottom margin*) to Niccolò Mannozzi, one of the friends or acquaintances to whom Galileo gifted copies, possibly to ensure his work reached specific places or networks where he aimed to clear his name. Courtesy of master & fellows Trinity College, Cambridge.

vital in retrieving, maintaining, and expanding Galileo's fama. After the trial had been completed, Galileo went on to publish an account of its proceedings (see fig. 1.2), which effectively broadened the trial's audience to include scholars from all over Europe, making them witness to his victory. Through a clever combination of tactics from the courtroom and from academic publishing, then, Galileo aimed to not just maintain his good name but even expand his fame.[53]

It all started with a manual Galileo had written around 1600 and that he sold, first in manuscript and from 1606 onward in print, to students and other interested parties. The manual accompanied an instrument Galileo had invented and developed a few years earlier with the artisan and instrument maker Marcantonio Mazzoleni and that he called the geometric and military compass.[54] To Galileo's dismay, he discovered in April 1607 that his former student Baldassare Capra had translated the manual into Latin from Galileo's original Tuscan. He had published this translation, with only very minor changes to its content, under his own name.[55] Adding insult to injury, Capra also implied that Galileo himself had committed plagiarism, purportedly having copied the instrument from a Flemish instrument maker.[56]

Galileo first heard about Capra's plagiarism from Giacomo Alvise Cornaro, a mutual friend of his and Capra. Upon receiving one of the first copies of Capra's book, Cornaro instantly recognized it as nothing more than a translation of Galileo's work. Cornaro lived near Galileo and immediately informed him, "not without indignant exclamations," of the fraud.[57] Galileo, shaken by what he had learned, set out for Venice as soon as possible. Once there, he presented his case before the Riformatori dello Studio on 9 April. He explained how he had designed and perfected a mathematical instrument, for which he had also written a manual. Now, he continued, the Milanese Baldassare Capra had translated his work into Latin and printed it in Padua, thus trying to "usurp his [Galileo's] work and honor." Galileo then explicitly petitioned the Riformatori to use their authority to restore his honor.[58]

Galileo's appeal to the Riformatori was the first step in the retrieval of his honor and his glory, which had implications for his finances and his scholarly credibility. Sales of the instrument and manual, along with private instructions for its use, provided an important supplement to Galileo's university salary, which he consistently deemed too low.[59] Capra's accusation of plagiarism now portrayed him, Mario Biagioli contends, as a lowly artisan rather than a "proper academician." If Capra succeeded in casting Galileo as a plagiarist, it would negatively impact Galileo's future credit among the students he instructed in the use of the manual. In this case, as Biagioli put it, "honor *was* money."[60] But the accusation and its implications for Galileo's honor were relevant to Galileo outside of the context of his private teaching as well. Like the fama legalis discussed in the first section of this chapter, honor was a prerequisite for full participation in society and in scholarship: Those without honor lost all credibility. Honor could be lost as a result of problematic behavior (if observed by others) or after being slandered, and it therefore needed to be carefully maintained and defended against slights.[61]

Capra's accusation of plagiarism cast Galileo as a liar and thief and thus threatened his honor. The problem was that Galileo could not convincingly defend himself against the accusations. It would be Capra's word against his, and with a plagiarism claim against him Galileo's word would not necessarily carry much weight. Instead, it would be much more convincing if a third party intervened. The institution most suited to this quasi-judicial task was the Riformatori dello Studio, which represented the university in its dealings with the Venetian civil government. The Riformatori were the singular authority that would intervene in university matters and furthermore had the power to intervene in publishing matters. Finally, their verdict would be above any suspicion of bias, as both parties in this dispute had ties to the university.[62]

Upon hearing Galileo's accusations, the three Riformatori first took steps to prevent the situation from escalating further: They instructed the studio's two presidents to forbid Capra's publisher and all the city's booksellers to sell any additional copies until they had investigated the case.[63] Then, on 19 April, they summoned Galileo and Capra to a meeting. Here they were each offered a chance to state their respective cases. Galileo repeated his allegations; Capra denied that he had accused Galileo of plagiarism, claimed that he had never seen Galileo's work in print, and asked that the prohibition on the sale of his work be lifted. He argued it was not a translation of Galileo's manual and offered much that was not to be found in Galileo's book. Capra also complained that Galileo had not taken up the proper arms of the scholar, namely pen and paper, and instead had turned to the authorities.[64] Galileo countered that an argument between two men of letters could indeed be solved through writing, but a conflict between a scholar on the one side and a slanderer on the other would best be dealt with in the present setting. He then demonstrated to the Riformatori that no reader of Capra's book would fail to understand the accusation leveled at him, and he offered them a paragraph-by-paragraph comparison of his work with Capra's, in order to prove that Capra's book did in fact plagiarize his own.[65] Capra replied that he had never intended to offend Galileo—but immediately thereafter accused Galileo of having plagiarized the instrument from someone else, namely a Flemish man named Johan Zugmesser.[66]

Witness testimonies provided the way out of this impasse. The Riformatori deemed Galileo's proposed comparison of the two works superfluous; after all, they already knew his opinion. Instead, they asked him to provide a comparison made by someone they trusted and whose expertise would be relevant to the matter at hand: Paolo Sarpi, the historian, mathematician, and theologian whom the Venetian senate had appointed to defend the government after Pope Paul V laid the entire city under Interdict. The defense was successful, and the Interdict was lifted right around the time Galileo needed his friend's help.[67] On 20 April Sarpi declared that almost everything in Capra's work had been copied from Galileo. He also swore that Galileo was the real inventor of the instrument: Galileo had demonstrated it to him ten years earlier.[68]

Before collecting Sarpi's statement, Galileo had already received the testimonies of two other witnesses: the aforementioned Cornaro, who had first discovered Capra's plagiarism, and Pompeo de' Conti di Panico. The date of their statements, 14 April, shows the speed with which Galileo and his supporters set to work to defy the accusations of plagiarism against Galileo.[69] Both witnesses declared that "some people," to "detract from his [Galileo's] fama" had accused Galileo of taking the design of his instrument from Zugmesser. This, they vouched, was not true: They had been present at an earlier meeting in which the two inventors had compared their inventions and had agreed that each had created his instrument on his own.[70] Like Sarpi, Cornaro and De' Conti di Panico were authoritative figures, and their value as witnesses rested not just on their being present at an earlier meeting but also on their social status. Cornaro was a Paduan nobleman who took a particular interest in mathematics and maintained a large network of like-minded acquaintances, and De' Conti di Panico, hailing from Bologna, had been welcomed into the Order of the Knights of Christ in 1598.[71] In their statements they took care to emphasize that they had not been alone at the meeting with Galileo and Zugmesser; scholars from several other nations had attended as well. They thereby suggested that their judgment of Galileo was shared more broadly. Perhaps they may also have expected that the Riformatori would see them as biased in Galileo's favor for patriotic reasons and thus preemptively stressed that many others with different origins shared their judgment.[72]

The accusation of plagiarism against Galileo refuted, the next step was to show that Capra lacked the mathematical insight necessary to have written the manual. This, the Riformatori decided after a suggestion by Galileo, should be done via an oral examination. They gave Capra a copy of Galileo's manual to study before their next meeting, which ended up being postponed because one of the Riformatori was too ill to attend. Given Galileo's eagerness to expose his opponent, one might expect him to be disappointed by this delay. Instead, he welcomed and praised it in his *Difesa*: Capra, now graced with more time to study his manual, could not claim that he had been given too little time to prepare.[73]

Galileo's next steps show that he had, at this point, started thinking about the long-term impact of Capra's accusations and how best to counter them. Of course, he still mainly aimed to convince the Riformatori that Capra was a fraud. But he also put forth a request to allow several gentlemen, "connoisseurs of the profession," to attend the oral examination. Galileo carefully selected his gentlemen-witnesses (Paolo Sarpi, Agostino da Mula, Sebastiano Veniero, and Antonio Santini) to include authoritative figures from Tuscany and Venice, the two most important territories during this stage of his career.[74] If they witnessed Capra's defeat, the four gentlemen-experts would be able to broker news of Galileo's victory beyond the milieu of the studio, or, as Galileo put it, they could "give, outside, an account of the conceit of him who has dared to publish me as the usurper, and himself as the true inventor of this work."[75]

At this point Capra tried to save face. When the meeting began, he offered to withdraw his work and proposed to write and publish an apologetic statement in the interests of restoring Galileo's fame (*fama*) and reputation (*riputazion*). Galileo refused this olive branch. Instead, he replied that his honor (*onor*) was in the good hands of the Riformatori (note that Galileo again specifically used the word *onor* here, as he did when first appealing to the Riformatori). He then moved on to interrogate Capra, in painstaking detail, about his mathematical skills. After this ordeal—its description in the *Difesa* fills nine whole pages—the Riformatori had finally heard enough and sent everyone home to await their decision.[76]

A few days later, on Friday, 4 May, their verdict was proclaimed "a suon di trombe" at the studio during its busiest hour. Students and professors alike heard that the Riformatori had judged Capra guilty of plagiarizing Galileo's work. In doing so, "he caused not a small scandal, and damage to the reputation of said Galilei, Lecturer in this discipline, and to the studio as well." The Riformatori therefore decided that his book could not be sold in Padua nor could additional copies be published.[77] This settled matters as far as the Riformatori were concerned. They had protected the reputation of the studio and the professor who stood accused of plagiarism among the audience of local scholars who had become aware of the scandal. To advertise their decision more broadly risked the affair becoming known beyond Padua. News of a student having plagiarized his professor would not be good publicity for the university, even if the matter had been dealt with appropriately.

Galileo, however, had both less to lose and more to win. The Riformatori's verdict restored his honor within the community of Paduan scholars, but Capra had also tried to usurp his "deserved glory." In contrast to honor, glory was not a prerequisite to participate in scholarship, nor was it something that people were generally presumed to possess. Instead, glory could be won: It was the recognition, bestowed by a wide audience, for special achievements. Although getting his glory back was not as crucial as restoring his honor, glory did yield benefits (such as the possibility to be considered for the Bolognese chair of mathematics).

Galileo would not let Capra get away with his attempt to steal what was rightfully his. The Riformatori's verdict, read only in Padua, did nothing to save Galileo's honor and glory abroad. This was especially problematic to Galileo, who claimed that by the time the Riformatori reached their verdict, thirty copies of Capra's book had been sold in "different parts of Europe."[78] To counter the potentially damaging effect of his adversary's work being available in far-flung places, Galileo decided to take up the arms of the scholar and write another book.

The *Difesa* is more than a personal defense against accusations of plagiarism; instead, it seeks to make its readers witnesses to Galileo's victory as lived in the courtroom. To that end, Galileo provided his readers with a detailed outline of his own clever exposing of Capra as a fraud. He also made sure to include the witness reports, thereby effectively making the work into a "manifestation of an authoritative network," as Nick Wilding has pointed out.[79] But the real verdict, as in any trial, came from the judges. Thus, at the end of the book readers could find a statement from the Paduan *podestà*, the city's chief magistrate, who confirmed that the facts cited by Galileo were accurate. Moreover, curious readers would be keen to know that he kept a copy of the verdict, *and* all testimonies cited, in his office, available for anyone to consult.[80]

The *Difesa* was included in the catalog for the Frankfurt Book Fair. This meant that it would come to the attention even of people who had never heard of the Paduan mathematician, his instrument, or the original manual—which, we may remember, was mostly distributed among Galileo's private students. Galileo's publisher seemed to have expected the book to sell well: He sent Galileo a sonnet thanking him for his beautiful gift, adding that the firm would be happy to print other works for him in the future. Galileo thus carefully exploited an unfortunate situation to work in his favor. As Wilding has convincingly argued, "By manufacturing a scandal and vehemently defending himself, [Galileo] not only protected his reputation but also made and disseminated it."[81]

Of course, reaching a large audience is not the same as convincing it, and Galileo certainly did not manage to convince all his new readers. The mathematician Mutio Oddi carefully avoided attributing the instrument to either Galileo or Capra in his treatise on the compass.[82] Moreover, a mathematician aiming to discredit Galileo in 1610, to whom we will return in the next chapter, claimed that among Venetians it was well known that Galileo had plagiarized the instrument known as the geometric and military compass.[83] Yet on the whole, the Riformatori's verdict and the *Difesa* seem to have been reasonably successful in achieving Galileo's aims. One of the letters Galileo received after publishing the *Difesa* stated that the Riformatori's sentence had chastised Capra before the people but that the *Difesa* had exposed him for what he was to the intelligent.[84] The Medici courtier Cipriano Saracinelli stated that Capra would certainly turn back the clock if he could: Galileo's pen had "castigated and whacked" him.[85] As late as 1620, Galileo received letters from people who wished to protect his reputation against Capra's slander.[86]

Capra's risky attempt to steal another man's thunder had failed miserably, and his reputation was thoroughly ruined by Galileo and his supporters. According to Cornaro, this was precisely in line with fame's fickle nature. Capra had intended to "make himself famous with the efforts of others," only to see his plan backfire, as "fame can change from the good and honorable, which one aspires to, into a vituperous plaintiff."[87]

Galileo's early reputation was shaped especially through the efforts of authoritative others, including friends and patrons. Aware of fama's value, they molded his reputation through talk and discussion, either spoken or written, in informal and formal settings, across a variety of contexts: a Florentine court case, several applications for university positions, and a highly volatile plagiarism conflict with a former student. Tracing Galileo's early reputation brings to light his struggles in the early years of his career to win the support of important patrons and advance his position. Though most of Galileo's contemporaries in these early years discussed him in a positive sense, there were elements to his reputation that were dubious at best and constituted a risk to his reliability as a witness and as a scholar. The accusations that he himself was guilty of plagiarism were particularly dangerous, and these continued to resonate when Galileo achieved wider fame, as we will see in the next chapter.

While underlining the importance of having a good reputation, this chapter has also shown that in specific situations, possessing this type of fama was not enough. To land an academic position, a different type of fama, namely widespread renown, could be a requirement. Galileo simply did not have this in 1588, but he started to work slowly on establishing a name for himself. He did so by inventing, by writing and publishing, and by fostering the sort of network that was expected of academic professors. However, strengthening his own position was not without risk, as Galileo discovered when he was quickly plagiarized after managing to establish a (modest) name for himself through work with his mathematical compass. But the risk of plagiarism was but one of fame's fickle rewards. Another was jealousy, shaped by competition for glory on various levels. The next chapter takes us to the period directly after the publication of Galileo's *Sidereus Nuncius*, showing how such competition structured responses to Galileo's breakthrough work.

2

Competition

GALILEO FIRST TURNED HIS TELESCOPE TOWARD THE HEAVENS in the autumn of 1609 but made his biggest discovery on 7 January 1610. That night he observed for the first time what he later realized were satellites orbiting Jupiter. Over the next few weeks Galileo continued his nightly explorations, quickly realizing that his findings were too remarkable to keep to himself for long. Although he first worked secretively and shared his news only with a limited few, he soon took the first steps toward publication, and the *Sidereus Nuncius* saw the light within three months of his first telescopic look at the night sky. The book, published on 13 March 1610, swiftly sold out. In the days and weeks after its appearance, queens, princes, diplomats, scholars, and gentlemen scrambled to get their hands on a copy of the work and, if possible, an instrument through which to observe Galileo's heavenly marvels for themselves.[1] But amid all the excited eagerness, there was skepticism as well. Could there really be "four new planets rolling about the sphere of Jupiter, besides many other unknown fixed stars," and was Galileo right that "the moon is not spherical, but endued with many prominences, and, which is of all the strangest, illuminated with the solar light by reflection from the body of the earth"? If true, the English diplomat Henry Wotton remarked, Galileo's claims would "overthrow all former astronomy."[2]

Galileo's findings indeed had far-reaching implications. Aristotelian philosophy firmly distinguished between the sublunar and the supralunar realms and held that the heavens, in contrast to the earthly realm, were unchangeable, incorruptible, and perfect.

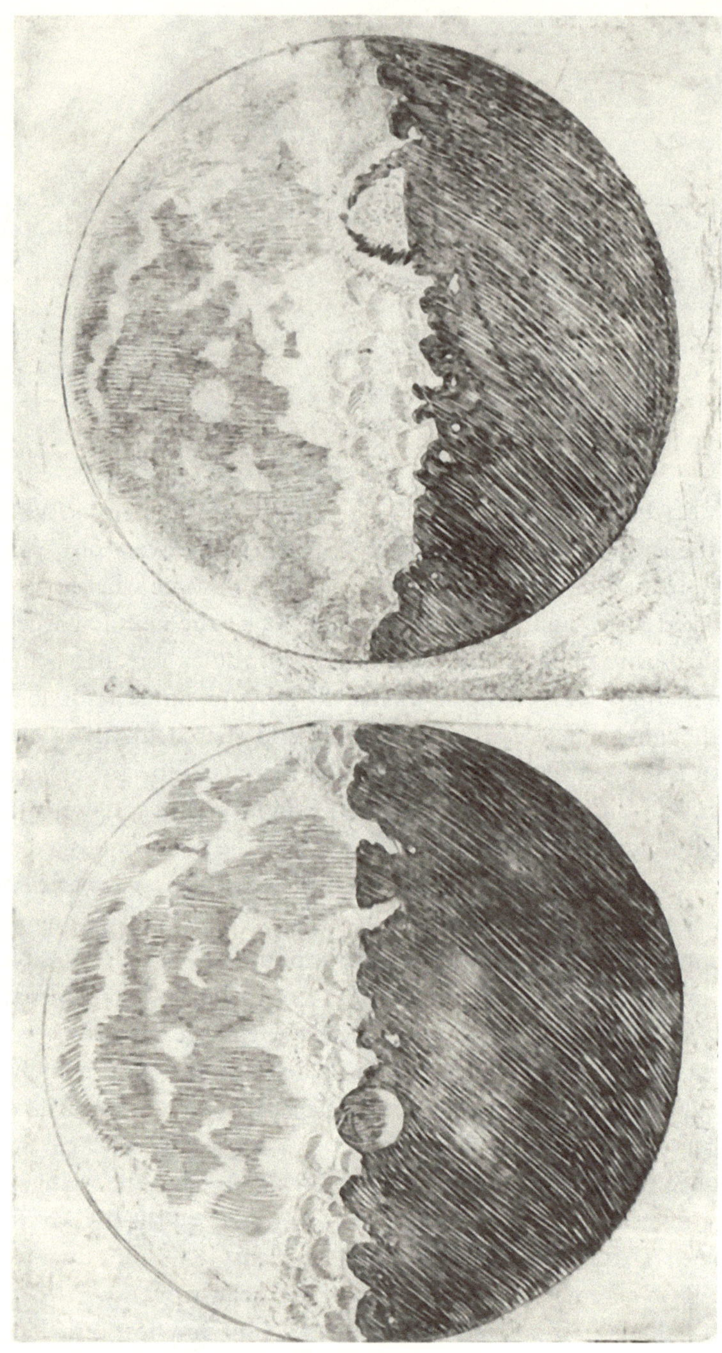

Fig. 2.1. Galileo's observations of the moon's rough surface, as printed in *Sidereus Nuncius*, 1610. Printed with permission from Cambridge University Library.

This tenet was in line with Christian doctrines that portrayed the earthly realm as the domain of humankind, subject to its sins and imperfections, and the heavens as the perfect realm of God. Aristotelian and Christian cosmology here were perfectly aligned, in part thanks to Catholicism's long-standing Thomist infusion with scholastic doctrines. However, if, as Galileo claimed, the moon was rugged and mountainous rather than smooth and perfect, could it really be said to differ from the earth? Jupiter's moons, too, posed a problem. That the earth was the only planet to have a satellite was a common objection leveled against the Copernican system by those defending geocentrism, and Jupiter's four moons removed it, proving that the universe had more than one center of motion (see figs. 2.1 and 2.2).[3] The discoveries' extensive ramifications prompted many of Galileo's contemporaries to approach them with caution, rather than embrace them immediately. If Galileo's claims were controversial, so was his method of discovery. The telescope—or *verrekijker*, *spy-glass*, or *occhiale*, as the instrument was initially known in 1611—had only been invented in 1608 and had so far mostly been used for military purposes.[4] Galileo, who had developed and improved the instrument since the summer of 1609, was among the first to aim it toward the sky.[5] Within Aristotelian philosophy the use of instruments to observe the world and thereby gain knowledge of it was highly contested, and some contemporaries refused to assign any validity to observations made through a telescope at all; Galileo's colleague Cesare Cremonini at the Paduan *studio* refused outright to look through the telescope, claiming that to do so would give him a headache.[6] Even more open-minded contemporaries frequently struggled to observe the phenomenon Galileo had described. The instruments he had developed were of a much higher quality than most others, and he carefully chose with whom he shared them.[7]

Galileo's name did not immediately inspire trust either, not least because many people simply did not know it yet. As we have seen, Galileo had previously published two other books, but one of those appeared under a pseudonym and the other was a defense against plagiarism—hardly the best way to establish a household name. Even Henry Wotton's long letter from Venice did not name Galileo but simply referred to him as the "Mathematical Professor at Padua."[8] In 1610, then, we find Galileo at the critical juncture

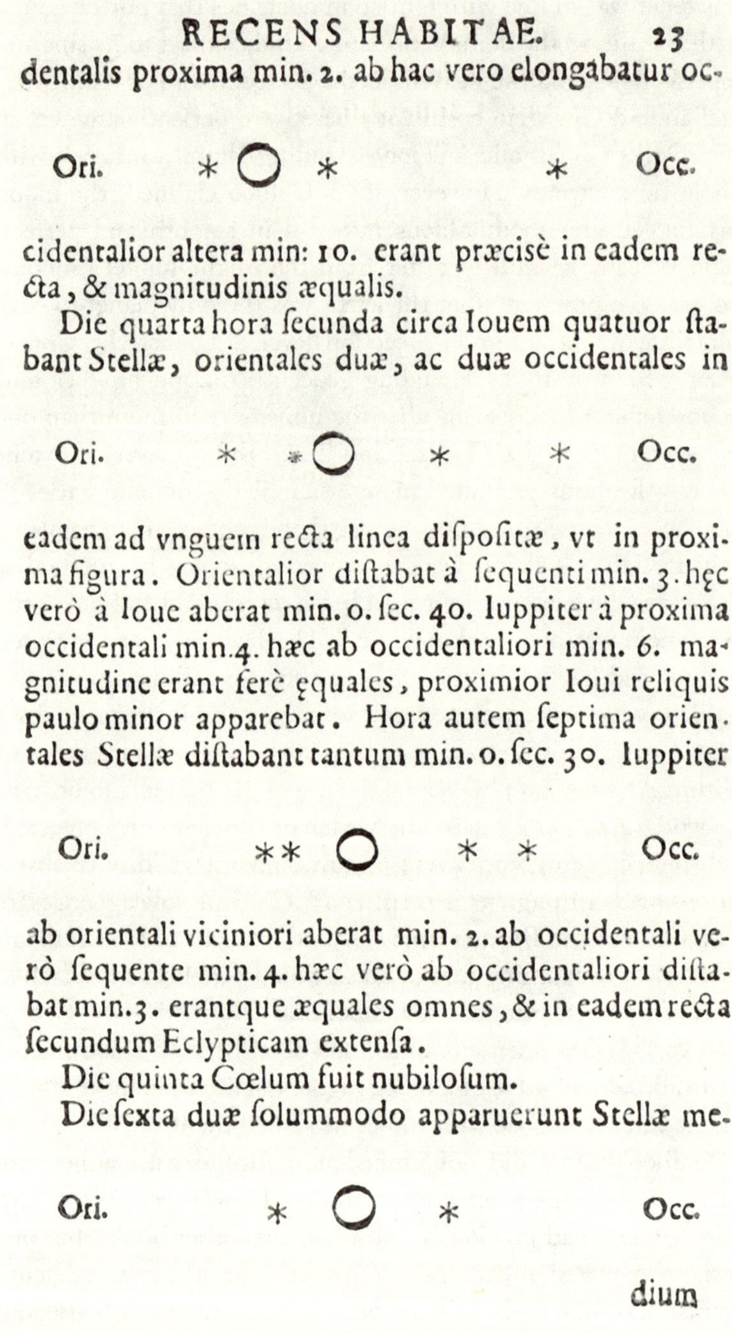

RECENS HABITAE. 23

dentalis proxima min. 2. ab hac vero elongabatur oc-

Ori. * O * * Occ.

cidentalior altera min: 10. erant præcisè in eadem recta, & magnitudinis æqualis.

Die quarta hora ſecunda circa Iouem quatuor ſtabant Stellæ, orientales duæ, ac duæ occidentales in

Ori. * * O * * Occ.

eadem ad vnguem recta linea diſpoſitæ, vt in proxima figura. Orientalior diſtabat à ſequenti min. 3. hęc verò à Ioue aberat min. 0. ſec. 40. Iuppiter à proxima occidentali min. 4. hæc ab occidentaliori min. 6. magnitudine erant ferè ęquales, proximior Ioui reliquis paulo minor apparebat. Hora autem ſeptima orientales Stellæ diſtabant tantum min. 0. ſec. 30. Iuppiter

Ori. ** O * * Occ.

ab orientali viciniori aberat min. 2. ab occidentali verò ſequente min. 4. hæc verò ab occidentaliori diſtabat min. 3. erantque æquales omnes, & in eadem recta ſecundum Eclypticam extenſa.

Die quinta Cœlum fuit nubiloſum.

Die ſexta duæ ſolummodo apparuerunt Stellæ me-

Ori. * O * Occ.

dium

Fig. 2.2. Galileo's observations of Jupiter's satellites, as printed in *Sidereus Nuncius*, 1610. Printed with permission from Cambridge University Library.

between local reputation and international renown, presenting us with the opportunity to trace the way the first shifts into the latter. Intriguingly, in the months after the *Sidereus Nuncius* came out and Galileo captured the attention of international audiences, his contemporaries frequently referred to him using a distinctly local identifier: his Florentine origins. In part, this followed Galileo's own self-fashioning on the title page of his work, which carefully situated him in two different regions: It listed Galileo's affiliation as a professor of mathematics in Padua and—in a font twice as large—identified him as a Florentine nobleman (see fig. 2.3). This self-identification was crucial for his attempts to win the patronage of the Medici family.[9] It also had unforeseen consequences for the reception of his work.

Early modern scholars often operated at a distance and needed to determine whether to trust one another without having met in person or without having had access to the necessary instrument to verify new discoveries or claims for themselves. Given these circumstances, scholars frequently relied on "strategies to stabilize knowing-at-a-distance."[10] Such strategies included tapping into local networks to gain detailed information about a scholar's day-to-day behavior and character. This practice intersected with newly forming ideas about regional (or protonational identities) as well as with a growing sense of competition. The imagined community of scholars spread out across Europe, the Republic of Letters, boasted ideals of cooperation and openness and respect, even—or especially—in light of differences in background among scholars. Because of this emphasis on open communication and shared, universalizing debate, the Republic of Letters has been studied as a public sphere *avant la lettre*, an idealized space in which competition and bias were frowned upon and not supposed to play a role in the way new scholarly findings were received.[11] Of course, ideals rarely correspond to reality, and scholars frequently fell prey to jealousy, impoliteness, and infighting. This chapter homes in on the interplay between the local, regional, and international networks through which Galileo's new fame was disseminated and negotiated. It shows the continued relevance of local settings as Galileo's newfound fame far transcended the more limited recognition he had so far enjoyed and highlights how fame put pressure on the ideals of equality. In particular, the chapter demonstrates

SIDEREVS

NVNCIVS

MAGNA, LONGEQVE ADMIRABILIA

Spectacula pandens, ſuſpiciendaque proponens

vnicuique, præſertim verò

PHILOSOPHIS, atq; ASTRONOMIS, quæ à

GALILEO GALILEO

PATRITIO FLORENTINO

Patauini Gymnaſij Publico Mathematico

PERSPICILLI

Nuper à ſe reperti beneficio ſunt obſeruata in LVNÆ FACIE, FIXIS INNVMERIS, LACTEO CIRCVLO, STELLIS NEBVLOSIS,

Apprime verò in

QVATVOR PLANETIS

Circa IOVIS Stellam diſparibus interuallis, atque periodis, celeritate mirabili circumuolutis; quos, nemini in hanc vſque diem cognitos, nouiſſimè Author depræhendit primus; atque

MEDICEA SIDERA

NVNCVPANDOS DECREVIT.

VENETIIS, Apud Thomam Baglionum. M D C X.

Superiorum Permiſſu, & Priuilegio.

Fig. 2.3. Title page of Galileo's *Sidereus Nuncius*, identifying him as a Florentine gentleman and Paduan mathematician. Printed with permission from Cambridge University Library.

how local and regional loyalties and rivalries shaped the reception of his ideas, revealing the crucial role of competition for fame as a driving force in the world of early modern scholarship.

JEALOUS VENETIANS, SHREWD FLORENTINES

The Republic of Letters formed a veritable treasure trove of information. Not everyone knew each other personally, but the community was small enough that a mutual acquaintance could often be found to share some insights into a scholar's behavior and character.[12] To those on the lookout for such insights, Galileo's provenance formed a useful point of departure, as it allowed them to tap into local networks of knowledge and reach out to friends they knew to live in the same city and who might be aware of local gossip and rumor. In other words, they were looking to get insights into the reputation he had within his community. As we have already seen, this type of *fama* was used in court cases to assess the reliability of witnesses, but it also came in handy when one needed to verify scholarly claims.

A prime example of this practice is the way Johannes Schreck, a German working in Rome, tried to get access to inside information that had not yet trickled down to Rome. Schreck wrote to his friend Giovanni Camillo Gloriosi, a Neapolitan mathematician who resided in Venice, where Galileo often visited friends and shopped for food and for supplies to make his telescopes.[13] Schreck hit the mark. Gloriosi not only knew Galileo personally—he had asked for his help when he considered moving from Naples to Venice in 1604—but he had also acquired a telescope and had even managed to observe two of the so-called Medicean stars.[14] Most importantly, the mathematician was eager to share local gossip about Galileo and the way his discoveries had been received there:

> Rumor has it that others already discovered two of them [i.e., Jupiter's satellites] by means of the spyglass. It is said publicly that Agostino da Mula, a Venetian aristocrat, was the first to observe these stars and that he informed Galileo, who knew nothing about them. His Excellency Messer Fugger told me he learned that the satellites of Jupiter were also observed by the Dutch, who invented the telescope.

> Perhaps Galileo was inspired by them and, to gain glory and money, even though he was not the first observer, he nevertheless wanted to be the first to write about them.[15]

Gloriosi's letter shows that in Venice, the rumor went around that Galileo had plagiarized another man's findings: Rather than being the first to observe Jupiter's satellites, he simply wrote down what someone else had discovered—a grave misstep. Gloriosi conveyed a second piece of local, public knowledge as well, indicating the lingering effect of Capra's plagiarism accusations. Everyone in Venice knew, Gloriosi told Schreck, that the geometric and military compass had been designed by the Antwerp mathematician Michel Coignet, but years earlier Galileo had presented it as his own invention. These pieces of local information and of the way Galileo was seen in a specific local community now found their way to new places, showing the continued importance of local reputation and rumor even as Galileo achieved wider fame.[16]

When Gloriosi sought to explain Galileo's wayward behavior to Schreck, he did so by pointing to his origins: "As you know the Florentines are shrewd and industrious: hence, grasping this propitious opportunity, he proclaimed himself author and inventor of the telescope and of the new planets, and received many honors and advantages from both the Republic of Venice and the Grand Duke of Tuscany."[17] Clearly, Gloriosi believed the satellites around Jupiter existed—he just doubted that Galileo had been the first to observe them. His plagiarism was entirely in line not just with his earlier behavior but also with the Florentines' "shrewd and industrious" character.

Gloriosi supposed Schreck knew about this Florentine character trait. That the Florentine reputation for shrewdness was more widely known is also suggested by another letter, this time from the French polymath Nicolas-Claude Fabri de Peiresc. Peiresc probably knew Galileo personally: in his student days the young Frenchman had spent some time in Padua, where he associated with Gian Vincenzo Pinelli's circle.[18] Peiresc, like Gloriosi, had verified Galileo's discoveries by using a telescope himself. He had even taken to observing the heavenly bodies closely and diligently, and by June 1611 he had managed to establish each individual satellite's orbit. He now thought that he deserved the opportunity to name this "discovery," as he had sufficiently deepened the understanding of Galileo's initial findings to almost transform them into something new.[19] Peiresc implied that his desire to rename the satellites could be seen as an infringement on Galileo's original discovery. But really, he wrote to his correspondent Giulio Pace, it was Galileo who had disappointed the Republic of Letters. Peiresc had simply stepped in as Galileo had failed to keep the promise he made in the *Sidereus Nuncius*, namely to publish more extensively on the movements of the individual satellites. Here, then, Peiresc invoked Galileo's origin: "Knowing that he is Florentine," he asks his correspondent, "can we dare hope for something other than what one would commonly expect from a shrewd person?"[20]

Gloriosi and Peiresc connected Galileo's Florentine provenance to his ethos. Highly trained in rhetoric, early modern scholars were used to making value judgments based not just on the validity of truth claims themselves but also on their authors' (supposed) reliability. They used various factors to do so. Among these were status and rank, as well as "outward signs (such as bearing, dress, and cleanliness)" and "inward attributes (such as demonstrated reliability, moral probity, experience, or even simplicity), and religious belief."[21] Together, these qualities said something about a scholar's ethos: the impression of character they left on their audience. Other classical appeals to ethos included sincerity, goodwill, and knowledge, while modesty, restraint, and the promise of novelty gained further weight during the early modern period.[22] Ethos could be conveyed in person or text, but Gloriosi's and Peiresc's letters show that Galileo's contemporaries also used his Florentine provenance as a stand-in to make claims about his character.

This fits with a broader development: Over the course of the sixteenth century, thinking about character—and especially national characteristics and stereotypes—had become more pronounced and articulate. This development related at least partly to the growing popularity in Europe of theatrical productions as inspired by Aristotelian principles of decorum: On the stage, it was necessary that the audience recognize each character as being driven by inner motivations that explained outward behavior. As theater became more internationally oriented and frequently featured characters from different parts of Europe, one element that was perceived to mark such inner motivation and the ensuing outward-directed behavior was provenance. Each nation was thought to possess different characteristics that distinguished its citizens from those of other nations, and such thinking about varied national characteristics gradually became more prominent and explicit.[23] Works on theater began to include lists detailing the precise character of every nation in Europe, spurring the development of easily recognizable national stereotypes.[24]

Such lists show a marked similarity to Gloriosi and Peiresc's portrayal of Galileo: they characterized Italians as shrewd, wicked, and deceitful. Take, for instance, the description by Julius Caesar Scaliger (himself Italian), included in his *Poetices libri septem* (1561): "One nation surpasses [the Spanish] in wickedness and is even less magnanimous: the Italians."[25] Almost a century later, the French author Jules de La Mesnardière wrote in his *Poétique* (1640) that "the Italians are idle, impious, seditious, mistrustful, deceitful and stay-at-home, subtle, courteous, vengeful, they love good manners and adore profit."[26] Of course, the political organization of Italy in the seventeenth century, with its various independently ruled regions and their distinct dialects and traditions, meant that local identities and loyalties often took precedence over a shared sense of "Italianness," and so it is not surprising to see that besides lists of national stereotypes, regional characterizations also circulated. At the same time, Italians from different regions sometimes rallied together if they felt attacked by a common opponent, especially if that person was from across the Alps. Scholars were especially used to thinking along these lines, as students at universities were organized in *nationes*: Bologna first divided students into two large corporations, the *oltramontana* for students from across the Alps and the *citramontana* for those below them, and then into smaller nations. Paris only recognized four nations (the French, Picard, Norman, and English-German), but Padua had at least sixteen, distinguishing between eight non-Italian groups, among them the English, German, Polish, and Scottish, as well as eight Italian ones.[27]

Besides the idea that Florentines (or Italians) were shrewd, other notions existed as well. Stereotypes regarding national character almost always consisted of both positive and negative attributes. On stage and in these lists, Italians were portrayed as refined, smart, inventive people who were sensitive to and appreciative of culture and the arts; as the list by La Mesnardière cited above indicates, Italians could also be depicted as courteous and cultured. Whether an author chose the positive or the negative image depended on various factors: context, the origin and point of view of the author, and power relations.[28] Galileo's supporters, as we will see, made eager use of the positive stereotype, pointing to the Florentine tradition for greatness to argue for Galileo's credibility. Gloriosi and Peiresc's selection of the negative stereotype was fueled by regional and protonational competition. Peiresc hoped to honor and to bring glory to France by renaming the satellites, whereas Gloriosi sought to defend the glory of the Venetian republic.

The Venetian response to Galileo's discovery was heavily shaped by his handling of the telescope on the one hand and his decision to move to Florence on the other. In the autumn of 1609, Galileo had presented his instrument to the Venetian government. His demonstration led to much delight and appreciation, and they had rewarded him by giving him a raise and appointing him professor for life, with an annual salary of 1,000 florins.[29] By the spring of 1610, however, the instrument had become much more common, and as a result the Venetians felt swindled. Giovanni Bartoli, the secretary to the Florentine ambassador in Venice, wrote to the Tuscan court on 27 March to say that in Venice, Galileo's book was being "read and considered by all," and the findings Galileo discussed in his work presented a "delightful pastime to the professors of these sciences." However, Bartoli added that appreciation for Galileo's discoveries was now mingled with a sense that others had been outwitted: "I cannot refrain from saying that many of these men now consider him to have mocked them, when he gave them this spyglass as a secret which was very common, and which is sold on the squares for even four or five *lira*, for the same quality, as is said; and many then laughed, calling themselves hasty, while he has tried to do his own thing, as he has done, and has succeeded in it with an augmentation of 500 florins to his normal salary for his lectureship."[30]

These initial irritations were aggravated by Galileo's attempts to move away from Venice. Galileo dedicated his astronomical findings to the Florentine grand duke, using the *Sidereus Nuncius* to enter into a patronage relationship with the Medici family.[31] By the time Gloriosi wrote to Schreck, at the end of May, Galileo's negotiations with the Florentine court were nearing completion: The Tuscan secretary had confirmed the grand duke's commitment to appoint him on 22 May and let him know the grand duke went along with his requested salary and title on 5 June.[32] The details of the negotiation were most likely not widely known, but Gloriosi's letter indicates that Venetians were aware that Galileo's discoveries had brought him, or would bring him, "honors and advantages" from the grand duke and vice versa. As a result, appreciation for Galileo's discoveries in Venice gradually curdled into an attitude of offended pride, along with envy directed toward Florence: Galileo, it was believed, had revealed himself to be ungrateful when he left the city to take up a position at the Florentine court.[33] Moving to Florence, Galileo would take with him the glory that—in the eyes of Venetians—belonged to the Venetian republic. In an attempt to hold onto it, Gloriosi claimed that the priority of discovery belonged not to Galileo and Florence but to the Venetian nobleman Agostino da Mula—who, attentive readers might remember, had helped defend Galileo against Capra's accusations of plagiarism in 1607.

The Venetians held on to this attitude of hurt pride for quite a while, as a letter from Gianfrancesco Sagredo, one of Galileo's closest friends, suggests. Sagredo had always respected Galileo's move to Florence, though he lamented his friend's choice when he first heard of it, writing that Galileo, abandoning the one city where he had enjoyed true freedom, had now embarked on "uncertain and dubious adventures" by returning to Florence. It was true, wrote Sagredo, that Galileo was now back in his home country, his *patria*, and could serve his natural prince, yet it was equally true that he "had left the place that had his best interest at heart." Sagredo also confessed his particular worry about the Jesuits' influence in Florence: "But this being in the place where the authority of the friends of Berlinzone, as we call them, applies, troubles me much, much more."[34] The phrase "the friends of Berlinzone" referred to a practical joke Sagredo had played on the Jesuits some years earlier, when he—pretending to be a rich old lady—had initiated correspondence with some Jesuits, one of whom was called Berlinzone.[35] Sagredo's anti-Jesuit stance was part of a more widespread feeling in Venice at the time. These years had come on the heels of the Venetian Interdict of 1606–1607, when Pope Paul V placed the city under an Interdict due to long-standing tensions with Rome. Originally concerning the autonomy of the Venetian clergy, the conflict had much wider implications: Disputes arose over the organization of the Paduan studio, and the role of the city's other schools, especially those of the Jesuits, was questioned. During the Interdict, the Jesuits were expelled from the city and would not be readmitted for several years, and a deep distrust of them due to their "meddling" was felt in the city.[36] Unlike Venice, Florence was known to be a city loyal to Rome, and from the Venetian point of view at least, Galileo was now settling in a place where the oppressive influence of Catholic conservatism was greatly felt. Galileo might have made his move to Florence precisely because he was vying for the approval of the Jesuits, but in the eyes of Sagredo at least, his relocation risked potentially severe consequences for his scholarly liberty and ultimately for the quality of his scholarship.[37]

It is therefore unsurprising that Sagredo welcomed the idea of Galileo returning to Padua. Few Venetians, however, agreed with him. Near the end of 1612, Sagredo wrote Galileo that talk had spread ("si è sparsa la fama") that he might return to the city and take up again a professorship there, as the Paduan studio was looking for a mathematics professor with experience and "great fame." Sagredo had tried to convince others that it would "be the best thing that could happen to the honor of the Studio" for Galileo to return. Yet while he found people who were willing to praise and esteem Galileo, he could not believe the "disgust" that others still displayed about Galileo's departure.[38] Earlier, Sagredo had told Galileo that several acquaintances no longer wished to be associated with Galileo and even wanted Sagredo to break off ties with him.[39]

The glory Galileo had won (and lost) in Padua played a role in the response of Giovanni Antonio Magini as well. Like so many other mathematicians, Magini—who had bested Galileo in the 1588 application for the chair of mathematics in Bologna—was heavily invested in finding out whether Galileo's claims held up. As he did not possess a telescope, however, he invited Galileo to his house to give a demonstration of the instrument. The demonstration took place in April 1610, but it had disappointing results. Galileo did observe two of the four satellites of Jupiter on 24 April and all four the evening after, but the others struggled with the new instrument.[40] Magini's assistant, Martin Horky, also managed to see all the satellites, but he still contested Galileo's claims. He even wrote to Johannes Kepler, saying that Galileo had a "terrible reputation" throughout Bologna ("tota in Bononia male audit"). He continued to paint a particularly grim picture of Galileo as deeply unhealthy, plagued by tremors, and a mess to look at.[41] Perhaps the months of hard work and nightly observations resulted in Galileo's messy appearance, and perhaps Horky was exaggerating, but the point is that information about the way Galileo was perceived in a specific local setting quickly found its way abroad. Magini's own letter to Kepler was less savage but similarly conveyed disappointment over the demonstration: "More than twenty of the most educated men were there, yet none of them could see the new planets distinctly."[42]

Magini then reached out to others and tried to convince them that Galileo was a fraud. News of this anti-Galileian campaign reached Galileo through one of his correspondents, Martin Hasdale, who informed him that among those Magini approached were the mathematician of the Elector of Cologne and mathematicians of "Germany, France, Flanders, Poland, England, etcetera."[43] Hasdale also thought he knew why Magini was so intent on discrediting Galileo. He was originally from Padua and especially resented Galileo for surpassing him "in his own patria; had it happened somewhere else, it would have hurt less."[44] Hasdale repeated this claim in another letter: Magini so vehemently rejected Galileo's claims because he found it difficult to endure someone else "stepping before him, especially in his own patria, in that profession where he wishes he was the only Phoenix."[45] Magini's jealousy was of a different nature than that of Gloriosi and other Venetians: He did not envy Galileo for his move away from Padua but for his initial success there. The jealousy may well have been fueled by the Paduan university's higher salaries and restrictions on hiring locally, which prevented Magini from joining the faculty.[46]

As Galileo's audience widened, then, the local setting where he spent most of his time did not lose relevance. On the contrary, his affiliation and provenance offered contemporaries useful clues regarding his credibility, while rumors about his supposed plagiarism that so far had remained largely restricted to Venice found a wider readership. Complaints about his ungrateful behavior, fueled by jealousy and regional competition over Galileo's newfound fame, similarly found their way to new audiences. Reputation and fame are often presented as starkly different assets, and it has recently been argued that personal details that situated authors within local settings became less relevant as audiences widened.[47] Yet the way Galileo's contemporaries tried to probe his credibility by turning to mutual acquaintances shows that the personal ties through which reputations are shaped and communicated were especially relevant at this critical juncture between Galileo's earlier reputation and his wider fame. The boundaries between the two forms of fama were porous at best.

RIVALRY AND SOCIABILITY

Competition did not just affect the reception of scholarly works on a regional level but also ran along protonational lines. Early modern scholars were well aware that this sort of rivalry and competition should, ideally, not play a role in the way they discussed their colleagues' work, but as Nick Wilding has recently pointed out, "Gentility was insisted upon so much because it was so lacking, not because it was universally accepted."[48] In practice, then, scholars not only frequently acknowledged the impact of rivalries but also discussed how best to balance scholarly ideals and the reality of competing interests, displaying a remarkable flexibility in the way they used the rhetoric of rivalry and distance to their own purpose.

We have already seen that Galileo's contemporaries used his origins as an argument against his credibility. His supporters, too, referred to his Florentine roots but in a wholly different manner: They connected them to a tradition of greatness to argue in his favor. For this, we turn to Galileo's fellow members of the Accademia dei Lincei and the publication of his 1613 *Sunspot Letters*.[49] This work brought together three letters written by Galileo that had previously circulated in manuscript form and that he wrote in response to Christoph Scheiner's work. Scheiner, a German Jesuit, published under the pseudonym "Apelles." The Jesuit order was normally quite open to new mathematical methods, and in 1610 many in Europe anxiously awaited their verdict about Galileo's discoveries. Though initially skeptical of his claims, the mathematicians of the Collegio Romano, led by Christopher Clavius, had by the end of the year confirmed most of his findings (the lunar surface remained the most controversial topic). Not long after that, however, the order's general sent out a memo to remind all Jesuit houses to strictly adhere to Thomist and Aristotelian teachings. Scheiner's work, which maintained that the sun was immaculate and unmoving and that the spots that could be seen to move over it were satellites—thus keeping damage to the Aristotelian view to a minimum—must be read within this context. His decision to publish anonymously, without gaining the permission of his order, should also be understood in the same light.[50]

Galileo and Scheiner each claimed to have priority of discovery and a better explanation for the sunspots, and the published collection of Galileo's letters was intended—by Galileo and the Lincei—to mark, once and for all, the superiority of his observations and understanding of the spots. To this end, the work would include many richly detailed illustrations, all paid for by the Lincei's leader, Federico Cesi. In addition to lending financial support, his fellow academicians provided Galileo with all the paratextual materials for the work and advised him, once they found out who hid behind the Apelles pseudonym, about the best way of attacking Scheiner. Attacking the Jesuits would only sharpen the order's campaign against Galileo, which he wanted to avoid. However, Galileo and his supporters were aware that the seemingly personal conflict between Galileo and Scheiner could be cast into broader terms by playing on a rivalry between Italians on the one hand and Germans on the other. Most of Galileo's letters about the matter are lost, but Cesi's replies suggest that he proposed attacking Scheiner's origins directly. Cesi, however, thought this unwise: "The liberty that you extend to me, gives me the courage to say that I do not think it would be good to attack the nation in any way, but better to deal an underhand blow to the person and the class. The nation is most hospitable to letters and scholars, and, with an abundance of books and editions, sustains their glory, and the Linceans specifically need to have their friendship: they are liberal in philosophizing, and I see that they honor the Italians greatly, when they do not have particular passion or envy." Cesi's arguments were echoed almost verbatim by Galileo's friend, the painter Ludovico Cigoli.[51]

After dissuading Galileo from attacking Scheiner's origins, Cesi added that he was "sure that your works will be esteemed there as they should be, and that they will have more honor than those of Apelles [Scheiner], even though he is from their nation."[52] This last comment demonstrates a keen awareness that scholarly works were not always assessed on merit alone and that national bias could affect the reception of new ideas.

Even if he was reluctant to attack the Germans outright, Cesi was not averse to putting this awareness to good use. The discussion soon turned to the "Letter to the Reader" that was to precede Galileo's main text. Galileo believed that the Lincei were being too aggressive in their promotion of him and asked that they adopt a more nuanced tone.[53] Cesi, in contrast, claimed that an aggressive tone was precisely what was needed. It was necessary to "rip away from the indifferent (who far outnumber your friends and adversaries together) those things sown by the jealous and other adversaries, who go and defraud you of your deeds." The problem was that "few are of sound and loyal mind; and of these, few in Germany, France, Flanders, and even here in nearby Naples, have a just account of the successes of the celestial discoveries. Your books have not gone everywhere: You have not printed everything. I can say for certain that many in those places have shown the things discovered by you; and if some of them have not dared to appropriate them completely, they still have not mentioned you." Cesi, in other words, believed that Galileo was not as well known in Germany as he thought he was, and they needed to seize this opportunity to counter the false accusations sown by people eager to appropriate Galileo's discoveries as their own. He was, however, willing to compromise. He proposed to add a poem on Florence instead: "But it is possible to make the preface more earnest, and one can with less affection and less show obtain the same effect. . . . Following the warning, the dedicatory letter will be slimmed down a bit. There was talk of adding an epigram in praise of Florence, to deal your adversaries an underhand punch."[54]

In the end, no poem was added, but the dedication, written by another Lynx and addressed to the Florentine nobleman Filippo Salviati, a longtime friend of Galileo, served the same purpose. Salviati was an appropriate dedicatee: he had been made a member of the Lincei in 1612 at the instigation of Galileo, and Galileo had carried out the majority of his observations on the sunspots at Salviati's villa just outside of Florence.[55] The dedication, written by yet another of the Lincei, emphasized the Florentine connection between recipient and author: "And what miracle is it if, in addition to knowledge of your merits, that bond of friendship with which he [Galileo] loves, admires, and esteems you, the Lynx, your native region, and your constant companionship also bound you together? It was with good reason that the noble city of Florence, so fertile with virtuous minds, extraordinary cradle of learning, which has always flourished and still flourishes in every virtue, should first taste and enjoy her own fruits and discoveries."[56]

With its emphasis on Florence as a cradle of learning, the dedication conspicuously placed both Galileo and Salviati in a long tradition of ingenious minds brought forth by the city. Foreign readers who did not recognize Galileo's name would here be reminded of Florence's status as a place of extraordinary scholarly achievement, which would, the Lincei hoped, make them more inclined to regard Galileo as a learned, cultured, and ingenious man. Praising Galileo's connection to Florence may not have violated the rules of scholarly sociability in quite the same way as attacking Scheiner because he was a German, but note that Cesi still used the phrase "an underhand punch" when he proposed to add an epigram in praise of Florence, indicating that finding the proper argument to defend an ally or attack an opponent was a fine balancing act that required both tact *and* a certain disregard for the unwritten rules of the Republic of Letters.

Ironically, the awareness that national bias also often played a role allowed for another appeal to provenance: to claim impartiality, one of the most esteemed features of the scholarly ethos held up by the Republic of Letters.[57] This shaped the way Johannes Kepler responded to Galileo's work. Kepler, imperial mathematician to Emperor Rudolf II, gave his opinion on Galileo's *Sidereus Nuncius* after being requested to do so by the Medici ambassador in Prague. Kepler willingly obliged. He first wrote a letter that circulated in manuscript form among his acquaintances and then, with slight revisions, was published under the title *Dissertatio cum Nuncio Sidereo* (A conversation with the sidereal messenger).[58] His was the first published response to the *Sidereus Nuncius*, and, strikingly, Kepler agreed to publish on the subject before he could personally verify Galileo's discoveries with a telescope. This heavily affected his discussion of the discoveries. His endorsement was grounded not on his own observations but instead on his assessment of whether the discoveries could indeed be true, based on the text and illustrations Galileo had provided and his trust in him as their discoverer.[59] It also affected his own reliability. Unable to simply offer an account of the procedure he had used to verify the discoveries, Kepler instead had to convince his audience that he was a reliable judge of Galileo's work, even though he lacked first-hand confirmation of the discoveries.[60] To this end, Kepler posed various rhetorical questions about why he should not "believe a most learned mathematician," whether he should "disparage him, a Florentine Gentleman," and if it would be "a trifling matter for him to mock the family of the Grand Dukes of Tuscany, and to attach the name of the Medici to figments of his imaginations, while he promises real planets?"[61]

Kepler also emphasized his own credentials as an assessor of Galileo's work at various points in the text. Especially relevant is the "Note to the Reader," where Kepler claimed his work had been criticized by friends who had read his *Dissertatio* and found his praise of Galileo excessive.[62] To defend himself against this criticism, Kepler again invoked Galileo's origins, now contrasting them to his own:

> Surely I have not colored my remarks about Galileo. I have always followed the practice of praising what I deemed well said by others, and of refuting what was badly said. I have never belittled or laid false claim to the ideas of others, when I had none of my own. I have never fawned on others, nor have I been self-effacing if I made some better or prior discovery by my own exertions. I do not think that Galileo, an Italian, has treated me, a German, so well that in my return I must flatter him, with injury to the truth or to my deepest convictions.[63]

Kepler here acknowledged the potential impact of national loyalties on the reception of scholarly claims, while he also made it clear that such loyalties did not play a role in his own verdict. The phrase had a second, subtler meaning as well. According to Martin Hasdale—the same person who informed Galileo about Magini's jealousy-infused campaign—Kepler greatly admired

Galileo's work but also felt that he had given German and Italian scholars cause for complaint, as he had omitted the names of other scholars who also contributed to the new knowledge he had laid out in his *Sidereus Nuncius*.[64] Kepler left this criticism implicit in the main text of the *Dissertatio*, although he did take care to situate Galileo's work against the background of works by earlier scholars, among them Nicolaus Copernicus, Giambattista Della Porta, Tycho Brahe, and Kepler himself. To the observant reader, however, Kepler's comment that Galileo had not treated him, a German, so well as to deserve his flattery might convey Kepler's superior impartiality.[65] Drawing on the tension between the ideals and reality in the Republic of Letters and the way competition for glory influenced the way new ideas traveled became a way to claim authority by positioning oneself as untouched by such concerns.

The remarkable flexibility with which early modern scholars approached the rules for scholarly sociability is further illustrated by Giovanni Magini, whom we have already met. Magini eventually came around to Galileo's side, not long after the latter received news of his appointment at the Florentine court. From that moment onward he suddenly emphasized their shared Italian roots, contrasting these with the German origins of one of Galileo's most vocal opponents, Martin Horky—Magini's former assistant.[66] However, when Horky responded to the *Sidereus Nuncius* with a virulently critical treatise, Magini tried to put some distance between himself and his former pupil.[67] After firing him, he wrote a mutual friend of his and Galileo's that this dismissal was meant "for all the Germans [*oltramontani*], who are the enemies of us, other Italians."[68] Letters from another of Magini and Galileo's mutual friends echoed this sentiment. Magini, he wrote, should not be blamed for Horky's inappropriate attacks: While Magini had tried to stop Horky before the publication of his *Peregrinatio*, he had failed because Horky was "very stubborn, as is the habit of the Germans" and the "people from beyond the Alps are always very eccentric individuals."[69]

Although the *Sidereus Nuncius* won him international fame that far transcended the local reputation he had previously enjoyed, the reception of Galileo's breakthrough work was still heavily shaped by the local settings in which he had worked before 1610. Paduan colleagues with inside information about his local reputation shared it with scholars in Florence and Rome, so that rumors and gossip about the lingering plagiarism accusations against Galileo found their way to new places and informed his fame there. Others invoked his origins as a sort of shorthand to stand in for more direct knowledge about his character. His Florentine roots proved a particularly flexible marker of credibility: Other mathematicians could argue for or against his credibility by either emphasizing its tradition of greatness or by pointing to the Florentine tendency to shrewdness. Their value and weight are underlined by the fact that the labels were used by scholars who had no access to either Galileo's book or a telescope themselves, as well as by those who did manage to verify Galileo's claims but nonetheless used his provenance to comment on his reliability.

The quick dissemination of Galileo's work across Europe brought out the best and the worst aspects of the Republic of Letters. We can observe the network of knowledge in full swing, as everyone set out at once to determine whether or not to believe Galileo's far-reaching claims, eagerly taking up pen and paper to ask distant correspondents what they had seen, read, and heard. In this situation, rumors abounded and information from local communities quickly found its way abroad, whether it concerned Galileo's own behavior in Venice or Bologna or the precise motivations Magini had for disbelieving and opposing Galileo. The way this sort of information was used to make claims about Galileo's credibility reflects the practices of the courtroom discussed in chapter 1 and shows that the reputation he enjoyed among the local communities in which he participated remained relevant as he gained wider fame.

However, Galileo's quick ascent to fame also exerted pressure on the Venetian republic's professed commitment to the ideals of cooperation and openness to triumph over cultural, religious, or political differences. Rather than conform to the image of a united, rational community, the republic fractured and revealed the various rivalries and loyalties within its boundaries. Jealous of his newfound success, many colleagues strove and schemed to maintain or obtain the same sort of recognition, displaying a blatant disregard for the norms of scholarly sociability in the process. Others were more cautious, seeking to bend rather than break the rules by dealing underhand punches. Still others, like Kepler, did uphold the republic's ideals and derived their position of authority from doing so. Crucially, however, all parties recognized the impact that individual, local, regional, and sometimes even protonational competition and rivalries had on the reception of new works and their authors, highlighting the continued relevance of local settings even as Galileo obtained widespread renown.

3

ADMIRATION AND APPROPRIATION

IN EARLY MODERN EUROPE, SCHOLARLY FAME WAS FASHIONED by ways and through means not immediately associated with the world of science we know today. The printed books that were the primary vehicle through which scholars communicated their novel findings were part of a much larger media system that also included poems, newsletters (*avvisi*), portraits, short biographies, eulogies, busts, and statues. These all played a role in the seventeenth-century economy of praise and glory, helping to disseminate new scholarship and the fame of individual scholars among scholarly and non-scholarly audiences alike. Commerce, fame, and good scholarship went hand in hand: booksellers keenly exploited the attraction famous names and faces exerted on potential buyers, while scholars helped promote each other's works by providing preliminary materials and praise. Such celebrations of great achievements served not only to praise individual scholars but also to signal what sort of behavior merited a following within the Republic of Letters.[1]

This chapter focuses on one specific genre that was not just highly relevant to the world of natural philosophers but also intimately tied it to the economy of praise and glory: poetry. Poets and poetry had long played an important role in the world of learning. As Crystal Hall has shown, they eagerly discussed "new, bizarre, or otherwise intriguing discoveries about the natural world," while verse had been an important "means for transmitting material with educational and informative value on a much larger scale" since Antiquity.[2] In the early seventeenth century, poetry also became an important means of transmitting praise to and bestowing glory on natural philosophers and their findings, thus pointing to the continued importance of cultural frameworks for the establishment of new ways of learning about the world. The humanist recovery of ancient ideas about fame and posterity meant that glory, in the Renaissance, increasingly became seen as the reward for particularly virtuous behavior. But to achieve eternal glory, one had to be remembered and praised by future generations, who in doing so kept alive the memory of great men. Historians and especially poets claimed an important role in this process of bestowing glory. Their writings, like those of ancient authors, would preserve a lasting record of the deeds and achievements of famous men: "Throughout Europe the poet was now to be celebrator of the great men of his own profession as well as those heroes of war and government who were his nominal subjects."[3] Poets, then, were powerful actors in the creation and dissemination of fame, and when in the early seventeenth century they increasingly wrote about novel scholarly findings and their discoverers, this was an important step in portraying the pursuit of natural knowledge as something virtuous. In particular, when poets admiringly discussed Galileo's achievements through imagery derived from ancient literature and myth, with references to heroic, adventurous conquest and battles fought and won, they helped bring natural philosophy into the realm of glory.

Among the most interesting authors and amplifiers of Galileo's glory were Medici court poets, a Jesuit poet and scholar in Rome, and Galileo's fellow members of the Accademia dei Lincei, who together will be the focus of this chapter. Although they were by no means the only poets to write about Galileo, their works allow us to explore a wide range of motives his contemporaries had for writing about Galileo. Writing and exchanging poetry was an important feature of the social life of cultural elites, grounded in the practices of courts, literary academies, and learned friendships. Before 1610, Galileo had already been the subject of a few poems, written largely within the context of the close friendships he formed in Venice, especially through the circle of Gian Vincenzo Pinelli, and, as noted in chapter 1, in 1607 Galileo received a short poem from his publisher to celebrate the publication of his *Difesa*.[4] After the appearance of his *Sidereus Nuncius*, however, there is a clear shift in the number and content of poems dedicated to him. The unique visibility of his achievements attracted new stakeholders to Galileo's name, which meant that poets could increase their own standing by being associated with him. They could moreover use his name in their poems as a sort of leverage, to acquire patronage, for instance, in attempts to develop new commercial interests or to promote new scholarship, scholarly ideals, and scholarly institutions. In other words, while most poets sincerely appreciated the marvelous nature of Galileo's discoveries, honoring him and his achievement was only one of their goals.

This affected their works and often entailed a reshaping of Galileo's fame in one or more ways as poets foregrounded one element of Galileo's story, left out another, or perhaps neglected to mention his name entirely to better fit a specific purpose. This points us to another mechanism crucial for the development of fame: Admiration and appropriation frequently went hand in hand. Such appropriations posed a risk to Galileo, who lost control over the narrative he had so carefully constructed in the *Sidereus Nuncius*.[5] But he was not the only one to run risks. The extraordinary nature of Galileo's achievement could easily overshadow any poet's celebration of it, and their contested nature made some poets cautious about offering their praise publicly, especially while the verdict on his *Sidereus Nuncius* was still out. In such cases, concerns about his discoveries also intersected with different ideas about the risks and rewards of print, as some poets actually favored manuscript exchanges. For the poets discussed here, all of whom were early adopters of Galileo's fame, writing about Galileo required a strategic calculation of the benefits and risks involved. In highlighting how they each balanced the prospective rewards and risks of investing in Galileo and his glory differently, depending on their closeness and connection to Galileo, as well as their personal interests and ambitions, this chapter reveals the often prosaic motives that lay behind this boost in celebrations of Galileo and his discoveries.

THE MEDICI POETS

Just a week after the *Sidereus Nuncius* appeared and sold out, Galileo wrote to Tuscan court secretary Belisario Vinta to tell him he was planning to publish a second edition in the local vernacular.

He also implied that several poets were jumping at the opportunity to contribute to this new edition: "The Tuscan Muses will not let this great opportunity to [celebrate] the glories of this most serene House pass, because already some are writing on this matter: and these compositions could precede the work."[6] However, Galileo's correspondence shows that the "Tuscan Muses" had not freely offered their services to him but that he had personally solicited them to write for him. He also seems to have refrained from telling them he wanted to print their works.[7] The poets, meanwhile, were not particularly forthcoming. As Mario Biagioli has shown, even when they did compose works to celebrate the satellites, their priority was to explore their meaning for the Medici family.[8] Here, I compare their hesitance to openly praise Galileo to the much more enthusiastic stance of Girolamo Magagnati, an old friend of Galileo.

Galileo's contact with the various court poets ran via an intermediary, Alessandro Sertini, who informed him of the progress Andrea Salvadori, Michelangelo Buonarroti the Younger, Piero de' Bardi, Niccolò Arrighetti, Gabriello Chiabrera, and Claudio Seripandi were making.[9] The poets all had ties to the Medici and to Florence's two most famous literary academies: the Accademia Fiorentina and the Accademia della Crusca. Both academies aimed to elevate the status of the Tuscan language and were sponsored by the Medici family; the Accademia Fiorentina in particular can be considered the cultural branch of the Medici regime.[10] Galileo was a member of the Accademia della Crusca and well acquainted with the Accademia Fiorentina, where he had delivered a lecture on the dimensions of Dante's hell in 1588. Still, the loyalty of the poets in this stage was clearly to the Medici family rather than to their fellow academy member. A letter by Sertini from 27 March captures his frustration and helplessness in trying to get the muses to write something: "The Muses are moving a bit slowly, as nine of them are waiting for a tenth to show some enthusiasm. If you want him to write something on the Medicean stars, you need to write him yourself. I have thrown in a word with Sig. Buonarroti; but with the others, not knowing them very well, I do not know what to do."[11]

The muses' caution relates to the controversial status of Galileo's claims at this point in time, as well as to the relative lack of public support he had so far received. Even as the *Sidereus Nuncius* had sold out immediately and sparked the interest of readers in Italy and beyond, by the end of March very few people had been able to verify Galileo's discoveries. Kepler had not even received his copy of the *Sidereus Nuncius*, let alone written an endorsement of it. And while the Medici were clearly interested in appointing Galileo as court mathematician, they had not yet made a final decision; they would do so only by the end of May. Galileo then wrote once more to further negotiate the point of his title—he asked to be called Mathematician and Philosopher to the Grand Duke—and the Medici granted his wish on 26 June.[12] Perhaps this was the sign the court poets had waited for: On the tenth of July, Galileo finally received word from Sertini that two poets had finished their works, one was still at work, another one had promised to get to it, and two would send him their compositions shortly.[13]

The same letter shows, however, that while they were now willing to write in support of Galileo, they were still on the fence about having their writings published. Sertini let him know that he did not "know what these men [Buonarroti and De' Bardi] think of using their names, in case you wish to print them . . . I will ask what they think."[14] In other circumstances, this hesitation to have poems printed might be interpreted as an extension of genre conventions. Court poets sometimes refused to publish, since print could be associated with a vulgar, commercial love of money and fame. Some authors therefore preferred to have their literary efforts read only in manuscript form, among a seemingly "aristocratic," limited number of friends.[15] In this case, however, there were other motives at play. In August, when Galileo had apparently pressed the matter, Sertini, somewhat annoyed, remarked that Galileo had "never said anything about printing" before. He added that he did not think the poets would want their names included: "Because you say you wish to print, everyone is afraid of it, and he [Michelangelo] does not yet want his name printed, but like Sig. Piero de' Bardi, should you print them, he would be content if it should say: of the Impastato, Accademico della Crusca."[16] This fear (*ogn'uno ne ha paura*) to have their names printed in one of Galileo's works was seemingly stoked by the still controversial nature of Galileo's discoveries. Although Galileo had secured his position at court by July, the same letter by Sertini opens with a brief discussion of the anti-Galilean writings of Martin Horky and Francesco Sizzi, which had just reached Florence and attracted much attention.[17] Such news also reached the city via other routes; just a week earlier, the Medici envoy in Venice had informed Cosimo that multiple writings against "the mathematician Galileo" had appeared.[18]

Given these circumstances, the poets were hesitant to wholeheartedly rally to Galileo's side, and the poems Galileo eventually received were not particularly spectacular. For the most part, they echo the *Sidereus Nuncius* in its emphasis on the nobility and immortality of heavenly bodies and their "natural" links to the Medici family. Buonarroti's poem, for instance, echoes his characterization of Ferdinando I as Jupiter and Ferdinando's sons as the four satellites surrounding him. These parallels to Galileo's own storytelling made it a highly suitable preliminary poem. Such poems appear before a book's *limen*—its doorstep or entrance—with the aim of preparing readers for their "entrance" into the main text and guiding them through it.[19] Preliminary poems have a reputation for being somewhat dull in terms of imagery and form, and the other poems Galileo solicited indeed share a highly similar repertoire of imagery and tropes. Buonarroti, Piero de' Bardi, and Niccolò Arrighetti all conjured up an image of the souls (*alme*) of the Medici princes finding eternal homes in the four satellites orbiting Jupiter, and their poems all praised Galileo's keen eyesight and his ingenious use of his instrument. Buonarroti referred to his "Lyncean gaze" (a reference to the Argonaut Lynceus), and De' Bardi and Arrighetti described the telescope as "illustrious glass" and "industrious glass."[20] The uniform character of the poems likely resulted from Sertini and Galileo's coordination. Buonarroti even expected Galileo to further change his poem in line with his personal preferences: "Sig. Buonarroti kisses your hands and sends you the attached composition, imploring you to improve them as you see fit, and [hoping] that it pleases you to welcome his goodwill to serve you."[21]

We will never know whether Galileo would eventually have managed to convince his muses to publish under their own names, as he never finished a Tuscan edition of the *Sidereus Nuncius*.[22] Still, De' Bardi's and Buonarroti's respective reluctance to have their works printed shows that even those who were relatively close to Galileo and stood to gain something by praising his discoveries—in this case, the approval of the Medici and the support of their newest client—were not necessarily keen on praising him openly and without reservation. Their loyalty to Galileo was limited, especially when compared to their fealty to the Medici family. This did change over time. As Galileo garnered the support of more and more people, the balance between risk and reward shifted. Buonarroti especially became a more enthusiastic supporter of Galileo after the summer of 1610, when he moved to Rome. In the following months, Buonarroti became part of a small group of Florentines who went around Rome to promote Galileo's cause amid *mormorazioni romanesche*, or "Roman murmurs," and he would later broker relations between Galileo and Maffeo Barberini.[23] The two men stayed in touch over the course of their lives, and Buonarroti remained a firm supporter of Galileo after the trial. They met in Florence as late as 1638, when Galileo had been placed under house arrest, and Buonarroti included Galileo in the gallery of famous Tuscans painted on the ceiling of the studio in the Casa Buonarroti.[24]

It is interesting to contrast Buonarroti's and De' Bardi's approaches with that of Andrea Salvadori, another Medici court poet. De' Bardi and Buonarroti were cautious and not especially forthcoming in their praise of Galileo, but Salvadori, who explicitly discussed the unbelief with which the Medici satellites had been met to that point, went so far as to omit all mention of Galileo. It was not for lack of space; the manuscript version contains twenty-three eleven-line stanzas. He also refrained from publishing the work during his lifetime, though he was usually quick to print his work; he most likely wrote his *Canzone di Andrea Salvadori per le stelle Medicee temerariamente oppugnate* in the summer of 1610, but it did not appear in print until 1688. Salvadori framed the conflict over the Medicean stars in terms of the Gigantomachy, the battle of the Olympian gods against the rebellious Titans. Setting up a contrast between the lowly, terrestrial nature of the challengers and the heavenly character of the Olympians, Salvadori characterized those people who did not believe the stars existed as arrogant, mad, and foolish liars.[25]

As Mario Biagioli has argued, Salvadori's poem legitimizes

the discoveries, but it does so only through the authority of the Medici.[26] In excluding their discoverer from the legitimation of the discoveries, Salvadori shows not only that his concern for the satellites rested entirely on their relation to the Medici but also that Galileo's name offered him no useful line of argument. Galileo tried his best to salvage the situation when he received the manuscript version, probably still hoping to include it in the Tuscan edition of his *Sidereus Nuncius*. He reworked Salvadori's poem so as to emphasize the durability of his discoveries and changed words and entire sentences to highlight their value to the House of Medici.[27] Still, even with this stronger emphasis on the utility of the discoveries for the Medici family, Galileo's name—the name of the discoverer who brought the family these marvelous monuments—remains absent, a prime example of the tension, discernible in all of the poems written by Galileo's "Tuscan muses," between Galileo's glory and the glory of the Medici family.

Not all Tuscan muses were as slow and hesitant as Buonarroti, De' Bardi, and Salvadori, however. The swiftest, most public, and most enthusiastic appraisal of Galileo's discoveries came from his longtime friend, the ambitious but struggling Girolamo Magagnati.[28] Their friendship went back to 1592, when they had met in Padua, and the two men exchanged letters, gifts of food, and visits until Magagnati's death in 1618. In the first years of their friendship, Magagnati worked as a successful and innovative glassmaker in Venice, where he was granted a fifteen-year privilege to make several kinds of colored glass and mirrors in 1595. By 1610, however, his situation was decidedly less rosy. Magagnati had spent several months in prison in 1602–1603, for, as he put it, helping a friend who in old age became besotted with a young girl.[29] His business also suffered. In 1604, his request for a twenty-five-year privilege to make a different kind of glass and mirrors was rejected, most likely after protests from the glassmakers' guild. He seems to have ignored the ruling, as the authorities destroyed the oven he had built in his own home not long after, and also rejected Magagnati's request for financial compensation. Finally, his application to renew his earlier privilege was initially granted, only to be severely limited after renewed protests from the guild.[30] Given these circumstances, it is unsurprising that around 1610 Magagnati was looking for a way to leave Venice and start over in

Tuscany. His letters show that he hoped to start a new company that would restore Tuscany's trade with the Levant and also extend to new territories, especially the Indies. To get this company off the ground, he aimed to enlist the support of the Medici family, in particular in the form of an investment of 10,000 ducats.[31] When Galileo published his *Sidereus Nuncius* and needed all the help he could get to boost his credibility and enhance his fame, Magagnati jumped at the occasion to help his friend and improve his own position. There is a clear parallel with the position of Giulio Strozzi, the Venetian poet who attempted to obtain Medici support by celebrating Galileo's discoveries a decade later, and it is telling that the most enthusiastic Venetian poems celebrating Galileo were written by people trying to move to Florence.[32]

Magagnati quickly set to work, and in May 1610, just weeks after the publication of the *Sidereus Nuncius*, he printed his *Meditazione poetica sopra i pianeti medicei.*[33] Both the visual framing of the work and its poetic contents show that Magagnati's main purpose was not to celebrate his friend Galileo but to flatter the Medici, especially Cosimo. The title page emphasizes the dedication to the grand duke and depicts the Medici coat of arms topped by a crown with five stars, one of which is larger and more clearly depicted than the four around it—a reference to the planet Jupiter and its newly discovered satellites.[34] Galileo's name is not on the title page, which further underscores that Magagnati's primary concern was flattering the Medici. This same prioritization of the Medici over Galileo is echoed in the text as well. Here, Magagnati stressed that under the family's rule Tuscany had far surpassed all other prosperous regions, including Rome, Lesbos, Thebes, and Crete, whether in the present or in the past. Now that Jupiter's satellites were named after the Medici, they finally had monuments appropriate to their glory, as Cosimo's name was now "printed in the celestial annals"—a far more durable medium than "rocks, metals and mountains, subjected to time, and dark oblivion."[35]

For all his praise of the Medici, Magagnati did not forget to credit Galileo's role in the discoveries of their marvelous monuments. Several lines of his long poem celebrated his friend for his mastery of nature and discovery of new orbs and new stars ("nov'Orbi e novi Lumi"). To Magagnati, this feat far surpassed the Argonauts' achievements or the simple introduction "of a new world to the world" by the Ligurian Columbus; note how regional competition again comes into play in this poetic celebration of Galileo's discoveries.[36] His choice of words and imagery echo Galileo's dedication of the *Sidereus Nuncius*, which similarly stressed the eternal nature of the stars, as well as Galileo's role in discovering them—making him the ultimate herald of the Medici family's glory.

In Magagnati's poem and the correspondence around it, we see how early modern friendship, patronage, commerce, and fame converge. While his praise for Galileo rested on sincere admiration and affection, there was an element of cunning to it as well. Offering poetry to a potential patron was a tried and proven method of gaining favor and benefits, whether material or immaterial.[37] Fame was a crucial ingredient in such exchanges. As Richard McCabe has pointed out, this was "a negotiation between the patron's celebrity and the poet's future fame—and only through the latter can the patron's memory endure."[38] Here, the negotiation between the Medici and Magagnati also revolved around the fame of a third party: Galileo. His discoveries were invaluable in their uniqueness, and Magagnati's personal connection to the astronomer gave him a further edge: Galileo would be able to send his work to the Medici family, and he would thus be acting as broker for Magagnati in the same way that Magagnati acted as fame broker for Galileo. Magagnati sent this poem to Cosimo as soon as it was printed, in May 1610, with a letter that emphasized his own urge to write about the planets and the glory they bestowed upon the Medici—he went so far as to say he had almost felt "forced" to do so.[39] Galileo sent Cosimo the poem, also in May, via his contact Belisario Vinta, making sure to mention Magagnati and noting that he was certain they remembered his name from earlier poems.[40]

The grand duke personally invited Magagnati to Florence but ultimately did not support his commercial plans; the endeavor Magagnati proposed was too risky.[41] Magagnati did reap other rewards from his labor, using the poem to extend and strengthen his network by sending it out to influential men in Italy and abroad, accompanied by letters that stressed his loyalty and willingness to be a worthy client. Among the recipients were Gabriello Chiabrera, a famous Medici poet residing in Savona, and the mathematician and diplomat Jacques Badovère, a close acquaintance of Galileo who resided in Paris. Magagnati asked him to forward a copy of his poem to Cardinal François le Joyeuse. Reading it, Magagnati hoped, would "lift for half an hour the weight of his most serious occupations" and perhaps briefly remind the cardinal of his "most humble servitude." Magagnati also sent his poem to the Accademia della Crusca and was rewarded by being admitted to the Florentine academy a few months later.[42]

In sending out his poem to these other figures and institutions, Magagnati did not have boosting Galileo's fame as his priority—he does not even mention him in the correspondence with Badovère or the Accademia della Crusca, where Magagnati emphasizes that the planets derive their glory from their dedicatee, not their discoverer. Still, the poem itself did praise Galileo, and it reminded readers of his role as the satellites' first discoverer. Such a public confirmation of Galileo's "ownership" of them was far from superfluous; we may remember that in May 1610, Gloriosi in Venice claimed that the Venetian Agostino da Mula should be credited for first discovering them. Magagnati's case thus shows how an ambitious author, one who was in close contact with Galileo, could seek to further his own career by capitalizing on and simultaneously enhancing the fame of Galileo's discoveries.

A Jesuit in Rome

In 1615, the poet, philosopher, and Jesuit Vincenzo Figliucci (1566–1622) published two poems on recent celestial discoveries in Rome. Figliucci seems to have had no particular relation to Galileo, although the two men possibly met in Rome in the spring of 1611, when Galileo, eager to win the approval of the Jesuit order, paid several visits to the Collegio Romano, where Figliucci was professor of moral philosophy.[43] Following a period of initial skepticism regarding his discoveries, the mathematicians of the college threw Galileo a celebration in the beginning of May, which led Galileo to believe he had won an important victory. He received more signs of recognition from the Jesuits around that same time: In the spring of 1611, a poem was read at the Jesuit college at La Flèche, and the verse discussed Jupiter's satellites and mentioned Galileo in its title.[44] Many of his friends remained on edge, however. The college's theologians continued to be wary of the possible implications of Galileo's discoveries, and shortly after the May celebrations the order's general sent out a memo to remind all Jesuit houses to strictly adhere to Thomist and Aristotelian teachings.[45]

The fraught atmosphere at the Collegio Romano in 1611 may account for Figliucci's decision to postpone publication of his first poem, which he wrote in 1611 but published in 1615. It may also explain why his two poems appeared under the pseudonym Lorenzo Salvi, although it is also possible that Figliucci avoided publishing under his own name because he wrote in Italian. Long before he published, Figliucci sent the first poem to several Sienese literati via an intermediary, asking for their opinion. They received it favorably and were curious to know the author's name, but Figliucci's intermediary told them that "vernacular poetry is not much practiced by the religious [i.e., members of the religious orders], and for that reason, as he is little versed, he does not care much to be known as author."[46] Note that the Jesuit Scheiner, Galileo's opponent in the sunspot controversy, also wrote under a pseudonym, and Figliucci's tentative attitude toward printing his work certainly fits in with the Jesuits' ambiguous attitude toward worldly glory, as discussed by Mordechai Feingold.[47]

The *Stanze sopra le stelle e macchie solari scoperte col nuovo Occhiale* brings together the two poems, a long dedication to the Aldobrandini family by Vincenzo's cousin Flaminio Figliucci, and two "brief explanations" in verse, in which Figliucci explains the contents, symbolism, and meaning of each poem in greater detail.[48] The poems have been characterized as outright celebrations of Galileo and indeed refer to him at various points.[49] Yet Figliucci did not necessarily write about Galileo in ways he would have appreciated. Instead, he shifted the narrative around Galileo's discoveries to fit his own goal of winning patronage and to celebrate the Jesuits' own contributions to the recent astronomical advancements.

Between the two poems, Figliucci's first one most explicitly praised Galileo and his discoveries. It opened with several verses on the telescope and then, in its thirteenth stanza, briefly but generously lauded Galileo. The stanza praised Galileo for having opened up new worlds that were previously not accessible to humankind, and it noted that his discoveries are particularly worthy because they concern the heavens rather than the lowly, muddy earth.[50] This emphasis on the distinction between these two domains kept intact precisely that element of Aristotelian philosophy that Galileo's telescopic observations undermined, and it is ironic that—precisely because it helped the poets praise Galileo—it would continue to be used in poetry for some time to come.[51] Figliucci's poem then moved on to the discovery of Jupiter's satellites. He praised them extensively in four octaves, situating them in the context of the court and presenting them as courtiers who orbit around their prince, taking turns in honoring him.[52] This imagery is almost identical to that used by Galileo and the Florentine court poets, but Figliucci did not mention the Medici family at all. By leaving out the Medici but keeping this imagery of ruler and courtiers, Figliucci hoped to flatter another family, the Aldobrandini, and obtain their patronage: In the dedication to Cardinal Pietro Aldobrandini, Figliucci stated that the celestial discoveries had been born under the same heaven as the six stars that adorned the Aldobrandini coat of arms.[53] Figliucci thus repurposed Jupiter's satellites, so clearly associated with the Medici family in Galileo's work and the poems of the Florentine courtiers, to win patronage from an entirely different noble family.

The bulk of the poem discussed other celestial discoveries, possibly in an attempt to draw attention away from the satellites that were so clearly associated with Galileo and the Medici and to shift the focus to discoveries in which the Jesuits, especially those of the Collegio Romano (whom Figliucci mentioned explicitly in the explanation following the poem), had played an important role. The many new stars Galileo discovered got eight octaves (twice as many as Jupiter's satellites!), and Figliucci also drew firm attention to the shapes around Saturn and the phases of Venus and discussed various hypotheses regarding the lunar surface. He concluded with a discussion, in verse, of the world systems of Copernicus and Tycho Brahe (as a Jesuit, Figliucci was bound to support the latter).[54] The second poem, which focused on the sunspot discoveries, even more clearly conveys Figliucci's desire to celebrate Jesuit contributions to scholarship. This time Figliucci was much more restrained in his praise for Galileo, which had everything to do with Galileo's conflict with the German Jesuit Christoph Scheiner.[55] As we have seen in the previous chapter, both men claimed to have been the first to observe the spots, and the conflict between the two men quickly became seen in the context of broader rivalries between Germany and Italy. In two stanzas that comment on this rivalry, Figliucci made it clear that he would not simply side with Galileo. Figliucci was originally from Siena, which had a long history of conflict with Florence, and any warm feelings Figliucci may have felt toward the Medici family were matched or even trumped by loyalties to the Jesuit order.[56] Figliucci consequently divided the credit for the sunspot observations evenly between his fellow Italian, Galileo, and his fellow Jesuit, Scheiner (who wrote under the pseudonym Apelles):

> But who, when admiring the Sun intently,
> Was the first to tell us of this news?
> Doubt still holds, and I do not flatter, nor lie,
> That the uncertain world until this hour names two.
> I hear proud Germany lauding you,
> Apelles, and celebrate this work, so beautiful.
> But Florence boasts of you, Galileo,
> And honors and sings your magnificent name.
>
> But of equal merit, and of equal praise it seems
> That both of them are worthy, to any who thinks
> As the one without the other dares to fix
> The invalid gaze in that immense light.
> One from the Rhine to the Donau makes them clear,
> And dispenses the beautiful secret to his Germans.
> One shows them to those here, from the Alps to us
> To the Arno, the Tiber, our Italian soul.[57]

Figliucci, while simultaneously praising Galileo and drawing attention to Jesuit contributions to scholarship, explicitly presented himself as an instrument of Galileo's fame. In the first poem he addresses Galileo explicitly—"you, Galileo, discovered the heaven to be rich with news stars"—and then also refers to his own role: "of which I speak in these songs."[58] However, his appropriation and reappraisal of Galileo's discoveries did not go over well with the Lincei, Galileo's most ardent supporters. Figliucci's poems circulated in manuscript before their publication and soon caught the attention of Federico Cesi, the Lincei's leader. Even before Figliucci's work was published, he warned Galileo that he had heard about a "modern poet" who had written about Jupiter's satellites but wished to dedicate his work to another prince, thus "using them in his own way without calling them Medicean."[59] And, when Cesi had finally read Figliucci's work (about a year later, in February 1615), he wrote Galileo that it "does discuss you, but not as much as it should, and it makes Apelles part of the discovery of the spots."[60]

Galileo's replies are unfortunately lost, but Cesi was not the only one who was offended by Figliucci's actions: In December of the same year, Luigi Maraffi informed Galileo that he had read the book. Maraffi was a member of the Dominican order, and it is possible his negative reading of the poem was informed by the rivalry between the Jesuits and Dominicans. He listed several points of critique that closely resembled Cesi's allegations: Figliucci had written in several places that the telescope was invented in Flanders and improved in Italy, but he did not say by whom; he wrote that others, including the mathematicians of the Collegio Romano, had observed Jupiter's satellites; that there were doubts as to who first saw the sunspots and that he, as an arbiter, must divide praise between Germany and Italy; and he never called Jupiter's satellites the Medicean stars.[61]

Galileo's supporters correctly interpreted Figliucci's poems and their purpose. His engagement with Galileo related not primarily to Galileo or his personal fame per se but reflects his hopes of using Galileo's famous discoveries to his own ends. As such, his poems emphasize an important consequence of fame: As Galileo's fame grew, he was discussed by people with whom he had no strong connection and who would appropriate his discoveries to serve their own goals. Those people may have further heightened his visibility by actively engaging with his work, but in the process they also reshaped his fame so as to better fit their personal aims.

THE LINCEI

Galileo, on his trip to Rome in 1611, did not just go to the Collegio Romano. The most productive outcome of his visit was likely his appointment as the sixth member of the Accademia dei Lincei in 1611. The academy had been set up in 1603 as a particularly exclusive institution by its founders, the Roman prince Federico Cesi and the Dutch physician Johannes van Heeck, who had initially asked only two others to join them in their new enterprise. Their goal was to promote the study of nature, especially through direct observation. They therefore chose the lynx as their emblem, evoking associations with the Argonaut Lynceus and the eponymous animal that was still to be found in Italy, both famous for their sharp eyesight.[62] The academy's first years were characterized by feverish productivity, as well as opposition and tragic loss. Van Heeck's story is especially gripping: He had been imprisoned for murder until Cesi got him out to launch his academy, but in 1604 Cesi's father, who opposed the enterprise from the beginning and was concerned both for his son's career and for his spiritual health, accused the Dutchman of heresy and denounced him to the Roman Inquisition. As Van Heeck had fled his home country because of his Catholicism, the accusation of heresy seems baseless. It is possible that Cesi's father simply found an accusation of heresy the most convenient ploy to get Van Heeck to stay away from his son, and that his allegations were actually fueled by suspicions of sexual intimacy between the two young men. The academy's rules stipulated that members should not take a wife, and early correspondence does seem to hint, at times, at relations that transcend friendship.[63] In any case, the allegations met their goal: Van Heeck fled Rome and spent several years abroad. He eventually returned but was plagued by bouts of mental illness, and his role in the academy was severely diminished. On top of this, the group's fourth member, Anastasio de Filiis, passed away in 1608, leaving the academy effectively with Cesi and Francesco Stelluti, the fourth original member.[64] To make up for the loss of De Filiis and the diminished role of Van Heeck, Cesi invited the inquisitive and innovative Neapolitan scholar and playwright Giambattista della Porta to join the group.[65]

The decision to appoint Della Porta resulted in the Lincei's initial suspicion of Galileo and his telescopic discoveries. The bonds between the academy's members were unusually strong, and they operated as a collective when one of them was under threat. When rumors of Galileo's work reached the Lincei in the summer of 1609 and the early months of 1610, they were mainly taken aback by his use of the new instrument; Della Porta had already theorized about such a device in his *Magia Naturalis* of 1589.[66] For more than a decade thereafter, artificial means to look beyond the horizon remained in the realm of legends, until a Dutch glassmaker in 1608 first managed to build a "spy-glass." Galileo, of course, heard about it in the summer of 1609 and set out to build an instrument of his own.[67] He never claimed to have invented the telescope, but this did not stop the Lincei from initially feeling a certain degree of resentment against him, and Stelluti wrote to one of his correspondents in September 1610 to say that "poor Galileo" would "be shamed," as Della Porta had already written about the telescope decades earlier. Shortly after that, however, Della Porta acknowledged that Galileo's remarkable contribution to scholarship surpassed his own.[68] When Galileo came to Rome in the spring of 1611, they invited him to demonstrate his instrument on the Janiculum hill, to their great delight and appreciation. Cesi asked Galileo to join the academy on the twenty-fifth of that month, and from that moment onward he shared in the academy's responsibilities and benefits.[69]

These were strongly regulated by the group's uncontested leader, Cesi. The young Marquis de Monticelli bore the title of *princeps*, established rules for the Lincei to live by, and assigned tasks to them.[70] These tasks served the academy's main mission of promoting the study of nature, and Cesi was adamant that the Lincei's efforts not remain restricted to Rome; in his vision, the Lincei would be known throughout all of Europe. The appointments of Della Porta in Naples and Galileo in Florence were important first steps, as were Van Heeck's European travels. But Cesi knew that the most important and effective medium of communication for the Lincei were books. Even when he sent Stelluti to visit buildings that might serve as the seat for a satellite academy in Naples, Cesi reminded him to keep in mind that such a building need not itself be famous or imposing. The academy's rules clearly stated that the "Lynxes are to have and obtain their name, honor and fame only with books and works."[71]

For the Lincei, then, fame was something honorable and worth striving for, and now, with the appointment of Galileo to their ranks, they had a particularly useful vehicle to further their own fame. Cesi had always expected the Lynxes to publish their findings under the academy's auspices. This usually meant that they included either a reference to themselves as a Lynx, printed the academy's symbol on the frontispiece of their work, or dedicated books to their fellow members and that the academy—or, more accurately, Cesi—contributed funds to cover printing costs.[72] With Galileo's appointment to the academy, this collaborative, communal practice of publishing further intensified as the Lincei used Galileo's recent publications to attract the attention of new audiences. Galileo's work on the sunspots was particularly suited to this purpose: His previous discoveries had already won him fame, and his work was therefore more likely to spark the interest of a wide readership than that of a lesser-known Lynx. Moreover, as it was written in response to the German Jesuit Christoph Scheiner, it was likely to attract the attention of German scholars specifically.[73] This was especially important to Cesi, who held German scholars in high esteem and hoped to gain their support as a next step in promoting the Lincei program beyond Rome and its immediate vicinity. As we have seen, however, the book's polemical nature also posed a problem: While it would attract new readers, those audiences would not necessarily be favorably disposed toward Galileo. As Cesi told him, he was less well known in Germany, and this might work to Scheiner's advantage.[74] The Lincei therefore took an especially keen interest in making sure this book presented Galileo and their academy in the best light possible. They provided the preliminary poems and other prefatory material for his *Istoria e dimostrazioni* (*Sunspot Letters*), took care of the practical side of the publishing process, and tried to manufacture the work's reception among various audiences.[75]

The three preliminary poems consist of two Latin epigrams and a sonnet in Italian, written by, respectively, Luca Valerio, professor of mathematics and philosophy in Rome; Johannes Faber, papal botanist and professor of botany, also in Rome; and Francesco Stelluti, jurist and one of the academy's original members. All three men were identified as members of the Accademia dei Lincei in the lines preceding their poems by the epithet "Linceo." All three poems explicitly comment on Galileo's fame to guide the reader to a positive reception of the work and to praise him for his masterful use of the "Lincean" method of studying nature. Together, they illustrate how the book became an instrument for the joint fame of Galileo and the Lincei.

Valerio's short epigram is the first of the three. It reminded readers of Galileo's earlier achievements:

When with your staff, Galileo, you saw the sky laid bare,
The startled Earth gave a strange cry,
For her brazen monuments must all yield to Time,
While you found lasting fame with a fragile glass.[76]

With its contrast between eternal, heavenly discoveries and earthly, transient monuments, Valerio's epigram is the most generic in theme and content; it plays on themes we have already seen in Galileo's work as well as in various other poems.[77] Still, Valerio's opposition of fragile glass and lasting fame is original, and it is noteworthy that Galileo here does not need to share his fame with a princely ruler or an opponent. This fits with the work's framing at large. Galileo here did not present new heavenly bodies named after the Medici family, and he did not dedicate this particular book to a member of the grand ducal family but to Filippo Salviati, a longtime friend who would later also be appointed Lynx.[78]

Following Valerio's reminder of Galileo's earlier achievements, the next poem introduced the topic of Galileo's new book on the sunspots. It firmly established Galileo's priority over Christoph Scheiner as their discoverer and, additionally, compared Galileo to two mythological figures:

You have no need, Galileo, of wings of Daedalian design
For flight and fall from a flaming Sun,
Nor will you, like some weak and unwarlike boy,
Sail above the stars astride Ganymede's eagle.
Splendid nature gave to you, as to the *LYNX*,
The eyes that breached heaven's walls:
With which you *FIRST* demonstrated the new stars in the firmament,
And showed new *SPOTS* are near the sun.[79]

Faber here commented on and legitimized Galileo's method of discovery by contrasting it to how two mythological figures came to "know" the heavens. The story of Daedalus and his son Icarus is familiar: To escape their prison in Crete, the architect and innovator Daedalus built wings out of feathers and wax. Icarus ignored his father's warnings and flew too close to the sun, so the wax melted, his wings fell apart, and he fell into the sea and drowned. The myth is closely associated with hubris, the ancient sin of arrogant pride, especially toward the gods. Faber's interpretation invokes a contrast between Daedalus's faulty "instrument," the wax-and-feather wings, and Galileo's modest but reliable one: his own eyes.[80]

The other ancient figure Faber invoked is Ganymede. This a much less obvious choice, as his story was not usually employed to warn against overconfidence. Ganymede, according to myth, was a beautiful young prince who was kidnapped by the ruler of the gods, Jupiter, who had taken the guise of an eagle. In early modern Europe, this myth was generally interpreted in one of two ways. The first situated the myth within a framework of learning and intellectual friendship. In this Neoplatonic version Jupiter is attracted to Ganymede's spiritual beauty, his pure and untarnished body and soul, and his name, which reflects his love for wisdom. The love between the two is transcendental: When Jupiter carries Ganymede to the heavens, the young boy's soul ascends from the earthly into the heavenly realm. An explicitly homoerotic version in which Jupiter's desire for Ganymede *is* sexual, not spiritual, also existed.[81] Faber's poem combines elements of the two interpretations. Ganymede "sailed above the stars" and thus came to know the heavenly sphere, but Faber also portrays him as a "weak and unwarlike boy." Again, the point is to praise Galileo's new, much more valuable method of discovery. Nature granted him and his colleagues ("as to the Lynx") an instrument that beat all previous ones, and their sharp eyes could even "breach heaven's walls."

The third and final poem, by Francesco Stelluti, brings home the Lincean message. His Italian sonnet presents Galileo as the master of nature:

> Your gifts, GALILEO, are now grown so great,
> That upon the torch of terrible Night
> A hundred Olympian peaks sprang up,
> Worthy thrones for your glory.
> And when you turned to them with your glass
> The highest heavens inclined in your favor:

ISTORIA
E DIMOSTRAZIONI
INTORNO ALLE MACCHIE SOLARI
E LORO ACCIDENTI
COMPRESE IN TRE LETTERE SCRITTE
ALL' ILLVSTRISSIMO SIGNOR
MARCO VELSERI LINCEO
DVVMVIRO D'AVGVSTA
CONSIGLIERO DI SVA MAESTA CESAREA
DAL SIGNOR
GALILEO GALILEI LINCEO
Nobil Fiorentino, Filosofo, e Matematico Primario del Serenifs.
D. COSIMO II. GRAN DVCA DI TOSCANA.
Si aggiungono nel fine le Lettere, e Difquifizioni del finto Apelle.

IN ROMA, Appreffo Giacomo Mafcardi. MDCXIII.
CON LICENZA DE' SVPERIORI.

Fig. 3.1. The title page of the *Sunspot Letters* book proudly identifies Galileo as a member of the Accademia dei Lincei. Printed with permission from Allard Pierson, Amsterdam: OTM: O 80-355.

Splendid stars were born by the thousand
Better to show your valor.
So the day-bearing Sun on lending you his light
Learns that you have long since seized it
From the brightest part of his fair face.
Thus you who showed us first
Those spots scattered about his bright mantle
Artfully adorn your name with a deathless splendor.[82]

Note the gradual buildup in the three poems and how they portray Galileo's fame as the rightful reward for his masterful conquest of nature. Valerio's short poem foregrounds Galileo's first celestial discoveries and praises the fame he has won through the telescope. Faber then goes further: Galileo's eyes have pierced the heavens, and this new method of discovery triumphs over old ones. Finally, in Stelluti's poem, Galileo's control over the heavens is so complete that they move, change, and sprout new peaks, thrones, and stars in honor of his glory.

This mastery of nature and the glory that results from it are not restricted to Galileo: By extension, all of the Lincei achieve them. The idea that Galileo best exemplifies an attitude toward knowledge and scholarship shared by all members of the academy also underlies the other paratextual elements. All of them work to identify or even brand Galileo as "Linceo" and make him the face of the academy.[83] Almost half of the title page is taken up by an image of a lynx, which is surrounded by a laurel wreath topped by a decorated crown (see fig. 3.1).[84] The connections between Galileo, the book, and the Lincei are further tightened as the title page stresses that the recipient of Galileo's letters, Marcus Welser, also belongs to their academy.[85] Galileo's and Welser's membership in the Lincei is given prominence over their other affiliations by printing the epithet "Linceo" in a larger font. The title text then features Welser's positions as a magistrate of Augsburg and an adviser to the Holy Roman emperor and Galileo's as philosopher and mathematician to the Grand Duke of Tuscany.[86]

Galileo's author portrait (see fig. 3.2) was a wholly Lincean enterprise as well. Just like the placement of a famous author's or patron's name on the title page of a work attracted readers, placing authorial portraits in scholarly works served to spark readers'

Fig. 3.2. Francesco Villamena's portrait of Galileo depicts him below two putti holding the instruments of his scholarly fame: his geometrical compass, the telescope, and the books he had published thus far. Printed with permission from Allard Pierson, Universiteit van Amsterdam, OTM: O 80-355.

interest and establish a sense of familiarity, as the image allowed the reader to catch a glimpse of what the author looked like.[87] The Lincei were well aware of this and took the initiative to add a portrait of Galileo by Francesco Villamena to the work—it was his first work to include one. They also paid for the design and worked closely with a friend of Galileo, the painter Ludovico Cardi da Cigoli, to develop it.[88] All this was not just for Galileo's benefit but for the academy's as well, as a letter from Stelluti shows: "All books are in the possession of our Librarian, who, when the due distribution to the Lincei and friends is completed, will have to deliver the rest, attaching the portrait in them to the benefit of the Company [i.e., the academy]."[89]

The image the Lincei hoped to present to the public through their most famous member was one of serene and unintimidating confidence. Galileo smiles benevolently at the reader and looks quite amiable, especially compared to Tintoretto's much surlier portrait of him from around 1605. He seems to have become gray in the years between the two portraits, and while his beard had changed very little, he somehow appears to have gained hair—an enviable feat. The changes to Galileo's person and status are also reflected in his clothing. In the Tintoretto portrait Galileo wore sober black, but this time he is in a fur-lined cloak, to reflect his membership in the Lincei: In 1606, the members wrote Emperor Rudolf II to inform him that they all wore robes trimmed with lynx fur during their gatherings.[90] Altogether, the portrait radiates an intimate yet effortless authority; perhaps the Lincei chose to show a nicer side of Galileo to balance the more polemical tone of the rest of the work. The imagery around the portrait, though, refers to earlier polemics in which Galileo's virtue triumphed over the envy of his opponents. As Eileen Reeves and Albert Van Helden have pointed out, the smiling mask above Galileo's portrait is crowned with a laurel wreath and signifies his virtue, while the one below him represents the envy of his opponents. The two *putti* to the left and right suggest that two specific (groups of) opponents are referred to here: The one on the left holds a compass, the instrument Galileo developed and sold in Padua (and which was copied by Baldassare Capra), while the one to the right holds the trumpet of the goddess Fama, not to the mouth but to the eye, referring of course to the instrument that has brought Galileo fame throughout Europe.[91] In this way the portrait's iconography reinforces the textual message of the *Sunspot Letters* and the dedication and letter to the reader (both provided by the Lincei librarian, Angelo de Filiis), which introduced Galileo's feud with Scheiner and emphasized Galileo's inevitable victory. Finally, the portrait also comments on the role of books as heralds of fame. The *putti* each hold, besides a scientific instrument, a book. These surely refer to *Difesa di Galileo Galilei contro alle calunnie & imposture di Baldassare Capra* (on the left) and *Sidereus Nuncius* (on the right) and underscore the Lincean idea that scholarly tools and books are equally important as instruments of fame.

In line with this notion, the Lincei did everything they could to guide the book's dissemination and positive reception. They sent the earliest copies to Welser, the addressee of Galileo's original letters, and to the bishop of Bamberg, who had resided in Rome but would shortly depart for Germany. The two men received the first copies as a sign of the Lincei group's gratitude. Commenting on the bishop, Cesi wrote that "he could help our cause and our name in Germany not in a small manner."[92] As a sign of goodwill, Cesi asked Galileo to meet the bishop—who would pass through Florence on his way back to German lands—and to tell him how highly Cesi and the other Lincei spoke of him and that they would all do their best to serve him.[93] Cesi also took care to ensure Welser's positive response to the work. At this point the Lincei had not done a full print run, partly because they had hurried the process along so as not to miss out on the opportunity of sending the work directly to Germany. However, Cesi also mentioned that they especially held back on printing more of the first folios, as Galileo still needed to comment on some elements of the prefatory materials. Possibly, Cesi also wanted to know how Welser would respond to De Filiis's preface, which fiercely attacked those who were too stubborn to base their conclusions on observation. In other words, De Filiis's text targeted Galileo's opponent Scheiner, and this put Cesi in a somewhat awkward position, as Scheiner too had addressed his letters to Welser, with whom he was in close contact. Before committing to a large print run, Cesi wished to reassure himself that Welser would not be offended too much by the more incendiary elements of the work.[94]

While being careful to micromanage the work's reception among its targeted, elite audience, the Lincei also took it upon themselves to disseminate it actively to as wide a readership as possible. Cesi initially considered printing 3,000 copies, but the large number of illustrations he and the other Lincei wanted to include drove up production costs to such heights that he had to settle for 1,400 copies.[95] This was still a considerable increase compared to the print run of the *Sidereus Nuncius*, which comprised only 550 copies, sold out almost immediately, and was quickly plagiarized in Frankfurt—a clear sign of high demand.[96] The Lincei apparently hoped that Galileo's newest work would match or even surpass this feat. They approached booksellers in Naples and asked Galileo to do the same in Florence.[97] In April, Stelluti sent Galileo twenty copies to distribute in Florence and informed him in the accompanying letter that a courier was on his way with another hundred copies. Ten of these, Stelluti told him, were different: They were printed "on finer paper, as *avviso*." In printing different versions of the work, the Lincei most likely hoped to attract different audiences, either by intending the text to serve as newsletters (*avvisi*) to be sent to European courts or as more popular reading that could be sold for a lower price. Of the hundred copies Galileo received, he was to give twenty to his friend and fellow Lynx Salviati, to whom the work was dedicated; the rest he could distribute according to his own wishes.[98]

Collaboration was a normal part of early modern publishing practices, but the extent to which the Lincei shaped the book and its reception was extraordinary. Their engagement with Galileo and his book even somewhat resembles the feverish enthusiasm of the Wittenberg publishers who, as Andrew Pettegree has shown, helped create "brand Luther." They, too, designed title pages, portraits, and paratextual material to create a marketable image of Luther, and in this process they sold the audience on the idea of Wittenberg as a publishing town as well.[99] The Linceans, albeit on a lesser scale, tried to do the same with Galileo. They made him the face of their academy and used his work to create awareness of the Lincei mission to promote a new way of studying nature. This way, they helped disseminate Galileo's fame among new audiences while also using him as a vehicle for their own glory. It worked, too. In 1614, a French student who visited the Jesuits at the Collegio Romano noted in his diary that according to the college's most prominent astronomer, Christopher Grienberger, Galileo was known in "all of Italy and Germany" as the "Lincean philosopher" (*philosophus Linceus*).[100]

The Lincei would remain involved in Galileo's career for years to come, keeping him informed about recent developments in Rome and giving him advice when needed. When Galileo published his next book, *Il Saggiatore* (*The Assayer*), in the autumn of 1623, they took a similar interest in the work and again helped with the practicalities of printing.[101] The timing was perfect. On 6 August the Tuscan Maffeo Barberini had been elected Pope Urban VIII, and the Lincei rejoiced: Barberini was known as a man of letters.[102] He had also corresponded with Galileo and in 1620 even dedicated a poem to him, which weaves in references to his astronomic discoveries but primarily discusses the dangers of flattery. The poem, titled "Adulatio Perniciosa" (Destructive flattery), is not only highly ironic in light of the pope's later condemnation of Galileo but also revealing of the ambiguities surrounding the pursuit for praise and glory in this period. It discusses the faith of kings, who are outwardly praised by their people and by courtiers happy to follow the glittering "glory of the royal scepter." But things are not always what they seem: "That which outwardly radiates brightness does not always shine within." The problem is that "truth avoids the dwellings of the powerful," and the king, "broken by defeat," will soon learn "what the courtiers' words of approval are worth." Those receiving praise, the poem warns, should not take it too literally: They may consider themselves lifted above everyone else, but the public's appreciation all too soon becomes poisonous: "How harmful and pleasant is the flattering tongue, exuding honey filled with hidden venom!"[103] Barberini may or may not have intended his poem as a direct commentary on Galileo's newfound glory, but his wary attitude toward slippery tongues and false words of approval is reminiscent of the suspicions harbored toward speech, especially by faceless crowds, we encountered in chapter 1. We will further pursue this topic, and its impact on Galileo's career, in the next chapter.

In 1623, however, none of the Lincei could expect Galileo's later struggles with the Barberini family, and they happily dedicated his new work to the new pope, letting Galileo know in a letter that they had already made the decision to do so. Galileo gracefully agreed.[104] A few weeks later Cesi presented the book to the new pope, and soon thereafter the pope's nephew, Cardinal Francesco Barberini, joined their ranks as a new member. These events were followed by a few happy years. In 1624, Galileo went to Rome, where he met most of his fellow academy members and, coached by

Cesi, had several meetings with the pope and his nephew.[105] While in Rome he also sat for portraits—both drawn and etched—by the painter and engraver Ottavio Leoni. A highly productive and fashionable portraitist, Leoni captured figures from Rome's political, intellectual, and cultural elite, and among the four hundred portrait drawings that were sold upon his death in 1630 are portraits of Urban VIII, Francesco Barberini, Michelangelo Merisi da Caravaggio, and Gianlorenzo Bernini. According to a contemporary of Leoni, Giovanni Baglione, his main criterion for drawing someone was their fame, as he chose to paint "Popes, Cardinals, and titled Lords, and [people] of every other quality, as long as they were famous, be they religious or secular."[106]

Besides the many, many drawings, Leoni also made a more limited series of about forty etched portraits. His etching of Galileo is similar to the earlier one by Villamena, included in the *Sunspot Letters* and reprinted in *Il Saggiatore*; it shows a slightly older and slightly sterner Galileo, but his hair, beard, and mustache are very similar, as is his clothing, with the exception of his fur coat (see fig. 3.3). However, while the Villamena portrait had been commissioned by the Lincei, this time it seems Leoni took the initiative to capture Galileo's likeness and include it in a printed volume with portraits of illustrious, virtuous men "of all the professions" he was planning to publish. Like the poetry of praise, the genre of illustrious lives was heavily indebted to humanist culture and practices of fame making, and authors usually discussed ancient authors alongside contemporary ones to underline the greatness of past and present and connections between them. Leoni, though, concentrated in particular on living, active subjects—painters and patrons of the arts foremost among them. In doing so he perhaps implied that contemporary artists deserved more recognition or fame than ancient authors did. He also claimed a role for himself as a maker of fame, in line with that of poets.[107] That Galileo is included in the drawings as well as in the etchings demonstrates his firm embedding in Rome's cultural scene as well as his distinct fame. Though Leoni also drew a portrait of Scheiner, Galileo was the only mathematician to be included alongside the artists, poets, writers, patrons, cardinals, and the pope in Leoni's collection of men who were *especially* praiseworthy.[108]

The volume, unfortunately, was never published; Leoni died

Fig. 3.3. Ottavio Leoni's portrait of Galileo, ca. 1624, shows him looking sternly ahead. Source: Rijksmuseum, Amsterdam.

in 1630, before he could finish it. That same year, Federico Cesi also suddenly died, and Stelluti and Cassiano dal Pozzo took over as leaders of the Lincei. Dal Pozzo would continue to play an important role in maintaining Galileo's fame—we will meet him again in the final chapter of this book—but with this change in leadership, the academy's glory years were in the past.[109] For some twenty years, though, Galileo and the Lynxes successfully collaborated, disseminating their fame and new scholarly ideals to mutual benefit.

The years after the publication of the *Sidereus Nuncius* saw a boom in poetic celebrations of Galileo and his works. The poems circulated as part of Galileo's books, as separate publications, or in manuscript form, but they were always in conversation with Galileo's writings. They lauded Galileo as a heroic discoverer of new worlds, exalted him for his sharp eyesight, praised him for his keen understanding of nature, revered him as herald to the Medici family, and celebrated the fame he had won through his discoveries.

There was a cumulative effect at play here. Many poets were initially quite hesitant to link their names to Galileo's, especially in print, a hesitancy reflecting both the initially contested nature of his discoveries and enduring ideas about the vulgarity of publishing. However, the more public statements of support Galileo received, the less risky it was to praise him and the more willing others were to add to it and link their names to his. We have also seen that not everyone shared the Florentine court poets' hesitations about publishing. The Lincei in particular enthusiastically and explicitly embraced books as instruments of fame and made full use of the possibilities offered by print to establish both Galileo's name and their own.

While the various poems disseminated and consolidated Galileo's fame, they also shaped and reformed it. Though most poets portrayed his fame as the direct and just result of his discovery of new heavenly worlds, their interpretations of Galileo's achievements were not always in line with the story he aimed to tell, and their appropriation of his work occasionally led to friction and frustration on his part and that of his supporters. The attitudes of the Lincei in particular reveal a sense of ownership over Galileo's public image that clashed with the way other people chose to engage with him and his work. It is a poignant reminder that the poets who bestowed praise on Galileo almost always had other motivations besides a sincere desire to praise him. That they hoped at the same time to gain something for themselves—patronage, prestige, help with a new investment, the advancement of one's own name and fame, status, and attention for a new scholarly academy—underlines the value and especially the collective nature of fame's rewards. These rewards, early modern brokers realized, extended not just to its recipient but could also benefit those who bestowed it, and this realization fueled a similar sense of ownership and competition that we have encountered in the previous chapter.

That the poets found ways to connect Galileo's cause to their own is not to say that Galileo did not benefit from their efforts. Even though most poets foregrounded elements that best fit their goals and left out things they found less useful, they still celebrated him as the discoverer of new stars and heavens. In doing so, they amplified his fame among new and existing audiences, but they also presented his fame as an entirely positive asset and his pursuit of novelties on par with those of ancient heroes. In this, they dramatically differed from the opponents we will encounter in the next chapter.v

4

Opposition

In the winter of 1614, on the fourth Sunday of Advent, Galileo became the subject of a fiery sermon in the Santa Maria Novella church in Florence. That day the hotheaded Dominican friar Tommaso Caccini discussed the tenth chapter of the Book of Joshua, in which Joshua successfully prays to God to stop the sun from moving so as to give the Israelites more time to defeat their enemy. Caccini later said he chose this topic not just to remind the congregation of God's almighty power but especially to scold "a certain opinion already of Niccolò Copernico, and in these times, as is very well known [*publichissima fama*] in the city of Florence, held and taught, as they say, by the mathematician Sig. Galileo Galilei."[1]

Caccini's actions formed part of a larger movement, and the sustained attacks he and his colleagues Niccolò Lorini and Raffaele delle Colombe launched on Galileo in this period coincided with criticism voiced by a group of academic opponents who would come to be known as the Pigeon League. Their opposition was a direct response to Galileo's growing fame and his enhanced visibility in Tuscany and in Europe. In the period under discussion, roughly 1611 to 1616, Galileo published new celestial findings as well as works outside of astronomy, circulated different manuscript letters, and became a hot topic of conversation in Florence's streets, squares, and churches. The pushback that followed not only tells us something about the specific resistance Galileo's work evoked but also highlights a mechanism inherent to fame: Galileo's widespread renown did not protect him from attacks but on the contrary opened him up to new criticism.

His opponents followed roughly two distinct lines of attack. The first targeted primarily Galileo's moral shortcomings and presented his supposed desire for glory as a sign of his vanity or pride, casting doubt on his integrity as a scholar. This claim was especially effective in the context of published works, which, because they were intended to reach a large audience, were easily associated with a thirst for public recognition. This argument also gained currency because of the novelty of Galileo's claims, which suggested he might be chasing novelty for novelty's sake, rather than out of a sincere concern for the truth. Suspicion of novelty also drove the second main line of argument, which portrayed his growing fame as dangerous and destabilizing because of the effect it might have on unwitting audiences, and it led the academics and the Dominicans to position themselves as protectors of their particular communities. This particular claim rested on a highly ambiguous view of the public as something in need of protection but also as a source of danger in itself, as crowds, mistakenly interpreting Galileo's fame as a sign of his merit, might repeat his novel and highly disturbing claims elsewhere, unchecked, without any form of control. Here we can trace how existing ideas about a particular aspect of early modern *fama*—namely talk, rumor, public opinion, or the chatter produced by large, uncontrolled crowds—found their way into debates on the value of scholarly fame.

The opposition Galileo's growing visibility evoked also highlights the continued relevance, in an age when publishing became more important for the establishment of fame, of oral and manuscript forms of communication. While Galileo and his opponents frequently referred to the differences between published works and talk, characterizing print as having lasting impact and talk as being ephemeral (for better or worse), the boundaries between the different spheres of communication remained porous. Fears about the effect of printed works were moreover rooted in anxieties about crowds that long surrounded talk; the main issue was the uncensored, uncontrolled spread of information, which meant it could reach anyone—including those unable to distinguish falsehoods from truths. In such a situation heresy might all too easily spread, and while Galileo's scholarly opponents especially targeted his printed works, the Dominicans were at least as concerned about the manuscript letters Galileo circulated and the loose talk that subsequently spread in the streets and squares of Florence.

While both the academic and the Dominican opponents objected to Galileo's fame on similar grounds, the dynamics of fame meant that their attacks had very different impacts. This especially shaped the possibilities for communication and retaliation he and his opponents had at their disposal, giving Galileo a clear advantage in the scholarly context but putting him at a distinct disadvantage in the religious one—and this would eventually have severe repercussions for his scholarship.

ACADEMIC ATTACKS

Galileo spent the summer of 1611 with his friend Filippo Salviati, who owned a villa to the northwest of Florence, in an attempt to escape the city's heat and unhealthy air. It was a well-deserved break from the busy period following the publication of his *Sidereus Nuncius*, Galileo's subsequent work on the phases of Venus, and his triumphant but exhausting trip to Rome.[2] But this retreat at Salviati's was not all play and no work. Salviati was a well-connected Florentine nobleman and frequently invited other guests to join his table as well. It so happened that one evening one of them, a professor of philosophy, found that he could not agree with Galileo on a crucial question: Why do some objects, such as a slab of ice, float on water, while others sink?[3]

Exchanges of knowledge in a social setting were an important feature of early modern scholarly culture and intersected with court culture, the academies, and the university. Disputes such as the one Galileo and his unknown opponent engaged in over dinner were rooted in the university disputation, in which elements of entertainment, play, and competition mingled. The goal was to win, and there were strict rules to facilitate the exchange, beginning with the distinct roles played by each of the three parties involved in disputations: The *respondens* defended a given proposition, the *opponens* tried their best to show it was false, and the *praeses* oversaw the procedure and proclaimed the winner. The competing parties tried to defeat the other by pointing out flaws in their reasoning. This way, students could practice the defense of canonic texts that formed the basis of the curriculum, which was mainly Aristotelian. Both the conflict and the outcome of the dispute ideally remained restricted to the event itself and did not follow participants to new places and later situations. This tenet existed to safeguard the participants' honor; they would not lose face in case they lost. In practice, though, the outcome of disputes often did transcend the narrow event itself. Winning in a spectacularly elegant manner could result in an extra reward, namely the audience's applause. The approval and glory won this way often proved a crucial asset for scholars, especially in the context of job applications.[4]

The dispute at Salviati's dinner party shared some affinities with these university disputations. At stake that first evening was, for Galileo and his original opponent, the validity of Aristotle's teachings in all matters philosophical. The value of the Aristotelian system lay in its closely knit, all-encompassing explanatory powers, and a disastrous domino effect would ensue if one of his principles proved to be false. Galileo's first opponent (we do not know his name; Galileo's notes do not mention it and none of his four opponents later claimed to have been present) defended the Aristotelian framework.[5] Aristotle held that water condensed by cold becomes ice; Galileo's opponent, from this, reasoned that it must have greater density than water and that the shape of a slab of ice must be what causes it to float. Galileo took the opposite position. Grounding his argument instead in Archimedes's principles, he claimed that ice was not condensed but rarefied water. The slab of ice floats, he held, because it possesses a lower density than water.[6] The debate became either so heated or so complex that no clear winner could be announced, and a new meeting was set up. This occasionally happened with university disputations too, and it did not necessarily pose a problem for the ideal that disputes should not follow participants to a new situation. It was crucial that participants continued to follow the rules of the dispute during and until the next meeting. Both Galileo and his later opponents would eventually break this tenet—and accuse the other party of doing the same.

Our most important sources on the dispute are the five scholarly treatises that Galileo and four others wrote on the matter. Together with a first draft version of Galileo's treatise, these published works, while highly rhetorical and often conflicting, allow us to reconstruct the chain of events that followed and trace the dispute's development from a largely oral event to a published polemic. The dispute has been discussed extensively before, and from Mario Biagioli's work we have a clear idea of the impact of disciplinary tensions and patronage relations on the dispute.[7] Not yet clear, however, is the role fame played in the dispute's development. I will show here that fame fueled Galileo's move from oral discussions into print in May 1612, when he published his *Discorso intorno alle cose che stanno in su l'acqua o che in quella si muovono* (*Discourse on floating bodies*).[8] It also shaped the way his opponents responded to him. All four members of the Pigeon League, as they were called (a pun on the name of their most ardent member, Ludovico delle Colombe), were less famous than Galileo, and it is tempting to interpret their hostility as envy of his success.[9] This undoubtedly played a role, but the four men also showed themselves concerned for the Pisan university and its students. They saw in Galileo's fame a real danger to scholarship, and it consequently became the red thread in their responses.

As noted, the evening discussion did not yield a clear winner. According to Galileo's notes, the philosopher he originally disputed with left the table, only to return a few days later and tell Galileo that he had a friend who was willing to continue the discussion.[10] This new adversary would be the passionate and tenacious philosopher Ludovico delle Colombe. Delle Colombe had already written against Galileo's *Sidereus Nuncius* in 1611 and had, when Galileo did not reply, appealed to Christopher Clavius in Rome, who also remained unresponsive.[11] He seemed excited at another chance to engage in a verbal duel with Galileo, this time in person, and his treatise suggests that with his arrival the dispute took on a more formal character. Rules were set and two judges were appointed to declare a winner, in line with the idea that disputes should have a binary win-or-lose outcome. The judges, Francesco Nori and Filippo Arrighetti, were Florentine clergymen with connections to both Galileo and Delle Colombe, which ensured their impartiality.[12] While most of the dispute's organization clearly resembles the way university disputes were given form, the judges were given an additional task that departed from standard university practice: Each participant would try to defeat the other not only through reason but also through experiments, and the judges would decide which experiments were valid.

However, the judges never could deliver a final judgment, as no further disputing seems to have taken place. The biased and highly rhetorical accounts of Galileo and Delle Colombe suggest that both participants spent a fair amount of the subsequent meetings violating the rules they had agreed upon. According to Galileo's notes, Delle Colombe failed to attend all subsequent meetings they had agreed on, thereby forsaking the dispute altogether.[13] He also held that Delle Colombe went around "many public places in the city to show to large multitudes of people some of his balls and tablets, . . . to sing his triumph by saying he had convinced me, while he had not even spoken to me."[14] Galileo here accused Delle Colombe of cowardice and dishonesty and of breaking the rule that the conflict and outcome of a dispute remained restricted to its original setting. As Mario Biagioli has pointed out, Galileo did not publish these allegations but chose to omit the names of his opponents altogether—most likely in an attempt to signal their irrelevance.[15] This tension between the semiprivate settling of disputes and their public impact would also play a role in the responses to Galileo's treatise.

Delle Colombe denied Galileo's allegations and, in turn, said that Galileo was the one who broke the dispute's rules. Having invited Prince Giovanni de' Medici and "a group of noble *literati*" to their scheduled meeting, Delle Colombe was greatly disappointed when Galileo, upon his arrival, refused to dispute. Rather than to "carry out the experiments with the appropriate greatness of figure and quantity of the material," Galileo announced that he would publish a treatise on the matter.[16] This way, Delle Colombe wrote, he could try to save face. Galileo would surely lose the dispute under its original rules but in writing could try "to make others believe with words, what he could not show through the senses; since, by altering and adding, and by taking away from the agreements and from the truth, one can easily with false premises and assumptions get rid of the true conclusion."[17]

Delle Colombe was not wrong. The experiment he had designed showed that a little ball of ebony wood sank, whereas a chip made of the same material floated. Shape thus seemed crucial, and if Galileo would continue with the meeting, Delle Colombe and the earlier opponent would win the duel. Galileo's options were limited. He could try to avoid defeat by claiming, in front of Delle Colombe, the judges, the Medici prince, and the other *literati*, that the original question had not been whether an object placed *on* water would float but whether such an object would behave this way if placed *in* water. If the ebony chip became sufficiently wet, it sank, and therefore Galileo could argue that its shape did not dictate whether it sank or floated. The problem with this strategy was that the experiments had already been agreed upon, in the presence of judges, and Galileo's opponents could (and would) thus accuse him of reneging on his earlier promise. Galileo's second option was to try to reframe the original question in new, broader terms. Again, he could not do so within the agreed context of the dispute. The question had been set, his opponent was too invested in Aristotelian philosophy and its tools, and disputes were too focused on winning rather than on exploring new knowledge and questions. His third option was to leave the live, oral dispute for what it was and instead publish a work on the matter. This option was preferable for various reasons. As court mathematician and philosopher to the grand duke, Galileo, as Biagioli has shown, could not stand to lose to such unworthy opponents as Delle Colombe and his allies. He needed to save face, and in his *Discorso* he explicitly claimed to have made the move at his patron's request.[18] A published treatise would moreover allow for the more expansive reframing that he desired. Not only addressing why some bodies float on water while others sink, he could also discuss the value of his particular mathematical-philosophical approach more generally.[19]

Galileo's fame provided a third and previously overlooked factor to spur his move into print, as a letter from Ludovico Cigoli shows. Galileo's artist friend wrote him in late August 1611 to say that he was wasting his time with the Pigeon League and should move on to better things. Cigoli had left Florence for Rome and advised Galileo to cast his gaze elsewhere as well; he considered Galileo to be in a league of his own and advised that he should not sink to the Pigeons' level by remaining enmeshed in such a provincial squabble. Cigoli acknowledged that the Pigeons needed to be dealt with, but the painter also thought that Galileo should debate his opponents only once. Attacking a more famous scholar was a well-established if frowned upon way for young, unknown scholars to attract attention and establish a name for themselves in the Republic of Letters, and Cigoli warned Galileo not to fall into the trap.[20] "These little birds," Cigoli wrote, "only try to make a place for themselves, not through their own valor but through their choice of rival." He also thought it was time to take the dispute public and satisfy his friends and his patron: Galileo should respond to his attackers not with "simple practices" but "mainly with good theories, so that they could not bite him as they are wont to do." Once he had dealt with the Pigeons, he could turn his attention to more worthy opponents who "are already famous, and known to the world."[21]

In his draft notes Galileo also presented the published book as the ideal medium for those dedicated to truth in the long run, contrasting the quiet stability of print to the clamorous instability of talk. "I gather that, as in the squares, in the churches, and in other public places, those voices that disagree with me are much more ubiquitous than those who agree with me," he wrote. These signs of opposition did not bother him, as "those who through their own carelessness persuade themselves to maintain falsehoods clamor more loudly, and make themselves heard more in public places, than those for whom the truth speaks, which, although it needs more time, calmly and silently unfolds and reveals itself."[22]

This specific defense of the printed book is especially interesting given the risks publishing entailed, showing us how the precise value of printed works was being renegotiated in this period. Published works allowed for a much larger and wider distribution, possibly among entirely new audiences, and as Nicole Howard has shown, this was precisely why many early modern scholars were wary of the printed book as a medium. Such wide circulation inherently meant a loss of control. Scholars especially feared possible corruption of their text and ideas by pirates and plagiarists, as well as misattributions and misinterpretations by readers unable to grasp the book's intended meaning—with potentially ruinous effects on the author's reputation and livelihood.[23] But the move into print could also evoke moral criticism from other scholars, and in this case it provided Galileo's opponents with rich ammunition for several dangerous lines of attack. They could now accuse him of refusing to dispute in person and of breaking the rules they had all collectively established earlier, thereby painting him as unreliable and a sore loser. Second, by going public with the dispute, Galileo committed the same faux pas of which he had accused Delle Colombe. He smashed the illusion that the outcome of disputes remained limited to those attending the event itself and considerably broadened its potential audience. That he published his work worsened the case against him. Because printed words could not be retracted—unlike easily forgotten verbal barbs—they could exert a long-lasting impact on reputations, and the "itch to write" was considered a scholarly vice when it led to invectives directed against other scholars.[24] Finally, Galileo's opponents could easily frame his move into print as proof of his thirst for glory.

The four men who together became known as the Pigeon League eagerly exploited these lines of attacks in the treatises they published between 1612 and 1613. All of them can be linked to the Pisan *studio* except for their leader, Ludovico delle Colombe. He instead had ties to the Dominican order in Florence—his brother, Raffaele, was one the clergymen who preached against Galileo from the pulpit.[25] Two other opponents, Giorgio Coresio and Vincenzo di Grazia, were or had been affiliated with the university. Coresio was professor of Greek until 1614–1615, and Di Grazia had taught philosophy until his term ended in 1610.[26] Di Grazia offers the most comprehensive refutation of the mathematical method as a legitimate means of finding truth, though his work is otherwise less aggressive and personal in attacking Galileo.[27] The most biting and personal criticism comes without a doubt from Delle Colombe's treatise.

The identity of the final member of the league, who was the first to respond but did so under the pseudonym "Accademico Incognito," is still debated.[28] Flaminio Papazzoni, the Pisan professor of philosophy, is Mario Biagioli and Stillman Drake's favorite candidate, while Michele Camerota and William Shea have instead argued for the studio's superintendent (*provveditore*), Arturo Pannochieschi de'Conti d'Elci.[29] I incline toward the latter interpretation, though it is also possible that the two men collaborated on this publication. The treatise includes a dedication by the superintendent's hand, in which the *provedditore* claimed that many people had asked him to translate the concerns of the "for now anonymous author," and it was "almost as if it was the task of the superintendent of the Pisan Studio to publish the defenses of others with regard to the doctrine practiced here, and taught by excellent philosophers, experienced and equipped for this purpose."[30]

The author's concern for the students at the university aligns with the particular role and responsibilities of the *provveditore*, as does the choice to publish—partly—under a pseudonym. The superintendent found himself in a particularly difficult position: One of the university's most visible professors had published a controversial work likely to draw attention to the studio, and Galileo's attacks on Aristotle—which implicitly also criticized the studio's curriculum—could not go unanswered. But how to intervene in this debate and distance the studio from its most famous professor without directly offending Galileo and his patron, the grand duke? The mask of the Accademico Incognito functioned as what James H. Johnson has described as a "manufactured buffer"—a shield that allowed the superintendent to address Galileo's criticisms and defend the other professors without escalating the conflict. Such an approach fits with the early modern appreciation for dissimulation as a useful tool to voice criticism: Hidden behind a mask or veil, one could make explicit things that would otherwise best be left unsaid, all while maintaining plausible deniability.[31] The work's playful tone and quality reinforced this buffer effect. The dedication is followed by a letter to a "Sig. Severo Giocondi," "Mr. Earnest Play," which stresses the importance of joking while studying and might, in case Galileo or his patron took offense, be used to deflect their concerns by claiming this was all a joke. Finally, the pseudonym allows the author to position himself—in marked contrast to Galileo—as appropriately modest and unguided by the desire for personal fame.

All four works share a strong dedication to Aristotle's teaching, a forceful rejection of Galileo's moral character, and a clear insistence that Galileo posed a threat to the Pisan student body. Galileo's fame connected these three issues and formed the starting point for both the Accademico Incognito's *Considerazioni* and Ludovico delle Colombe's treatise. Both men acknowledged that fame had positive features. The Accademico Incognito immediately foregrounded the attraction it exerted on readers, especially as it was often taken as a sign of merit, and admitted that he had been drawn to Galileo's work for these reasons: "I, because of the fama of this man and of his experiences and observations, eagerly sat down to read it."[32] Delle Colombe, in his opening statement, similarly stressed fame's affective appeal. He turned in particular to the attraction fame exerted on scholars and its ability to inspire people to virtuous deeds. Within the Republic of Letters, the idea that scholars could be inspired to work hard, behave virtuously, and serve the truth because in doing so they had the possibility of earning eternal glory was widely shared. Delle Colombe too acknowledged that scholars driven by this desire for recognition were not necessarily "blameworthy and greatly benefit from it."[33]

Yet, he also questioned whether this desire was entirely appropriate. Those who achieved "such glorious remembrance through new things" were, he wrote, "not just admired, but reputed to be Gods." Surely the desire to be admired or reputed as God was immodest, vain, and proud and not befitting a proper scholar. He added that very few men could ever hope to achieve the level of fame they aspired to: "As this particular talent is given to so few men, many, longing to enter the lists, because of the difficulty of the enterprise do not accomplish the desired outcome of [finding] truth."[34] In some cases, then, the desire for glory led scholars to develop scholarly vices. Wishing to see the exceptional nature of their work and erudition confirmed, they set out to elevate themselves rather than the Republic of Letters as a whole; the desire for individual glory thus took over from the more appropriate longing to serve the community. Such self-aggrandizement led to all kinds of unwanted behavior. Scholars might be tempted to publish sloppy work; they might choose to address an easy, fashionable, and useless topic rather than devote themselves to something more difficult and important; or they might attack other—often more famous—scholars to build a name for themselves. In extreme cases, scholars might even publish views and conclusions that they did not themselves believe but that would, they knew, attract attention.[35] This is the implicit accusation in Delle Colombe's opening statement: "There are those intellects, who, similar to these [the finders of new things], hope to make the same things seem new that have already been discarded on account of being false, deriding the same inventors, and who today, which shines as such a beautiful day of truth, want to overshadow others with the darkness of their intellect. . . . Do they want to joust the first writers of the world as if they are equals, without knowing what their adversaries' weapons are made of, and without having sharpened their own?"[36]

Delle Colombe suggested that Galileo attacked Aristotle "as if they were equals" out of a misplaced sense of ambition. Criticizing Aristotle might be allowed as an intellectual exercise, such as a disputation, but he accused Galileo of doing it out of a desire to make a name for himself. This led him to commit a serious scholarly vice, according to Delle Colombe: he was defending the ideas of Archimedes, even though he knew them to be false.[37] Galileo's work was a vain "exercise in ingenuity," characterized by inconsistent reasoning and many mistakes, some of them even "da voi fatto ad arte": Galileo was willing to contradict himself, Delle Colombe wrote, as long as he could gainsay Aristotle as well.[38] Indeed, the Florentine "in almost everything shows himself to oppose Aristotle, in which is the sum of philosophical truths, turning many of the old opinions into novelties." The image Delle Colombe conjured up was that of Galileo viciously attacking Aristotle as if he were a punching bag, without any moderation or consideration for truth.[39] Careful to avoid similar criticism, lest he be accused of excessively attacking Galileo, Delle Colombe foregrounded his loyalty to Aristotle: "Seeing that he has wrongly become an Antiperipatetic, in this particular thing I would like to become an Antigalilean, out of gratitude to that great prince of many academies."[40]

By stressing that scholars would try to find glory through "new things," Delle Colombe portrayed Galileo as a typical *novator*. Unlike today, in the early modern period scholars mostly viewed the search for novelty and innovation with suspicion and fear, especially when novelties were framed as contradicting Aristotelian philosophy.[41] *Novatores* mounted attacks on Aristotle without presenting an alternative authority and were therefore suspected of chasing novelty for novelty's sake. The term had its origins in the religious debates between Catholics and Protestants and could thus evoke associations with heresy, but Delle Colombe did not go so far as to accuse Galileo directly of such a thing.[42] It was enough to connect Galileo's novelty seeking to his desire for glory. The accusation was particularly effective given Galileo's situation. Had he not recently acquired fame through his discovery of new stars? Had he not used, for those discoveries, a new instrument that he had sought to make his own? And had Galileo not come by his fame very rapidly—too rapidly? Such a quick rise was often associated with the fickle kind of fame that did not follow true merit and virtue but rested on falsehoods and lies.[43] The association with falsehood also reared its head in connection with the telescope: after the publication of *Sidereus Nuncius*, several scholars had accused Galileo of presenting the *occhiale* as his own invention.[44] Delle Colombe did not rehash these criticisms, but among knowledgeable readers his opening sentences may have gained traction in light of them, especially as he now accused Galileo of reviving the ideas of Archimedes just to further his own glory. Galileo's newly won fame thus made him vulnerable to criticism: It allowed his opponents to frame him as someone ruled by a desire for glory, unsuited to the pursuit of knowledge, and a threat to those who adhered to the values of scholarship.

His opponents also portrayed Galileo's desire for glory and novelty as a potential menace to a more specific community: the young students at the university. Though he was not obliged to teach, Galileo's recent university appointment had given him an influential position. This worried his opponents, not just because they were wary of Galileo's intentions but also because they did not entirely trust the judgment of the students they claimed to serve. This objection against Galileo's fame reflects deep distrust of the origins of fame as formed by large groups of people without specific authority, or the ability to distinguish between proper, deserved glory and false fame. It is especially present in the texts of the Accademico Incognito, who presented Galileo's work as an attack on traditional ideas and education and positioned himself as protector of the university and its students. Who knew how many young men "of quick intellect and curious to know many things" would be "enticed by the novelty of the doctrine" and "recklessly stray from the straight and safe road of Peripatetic philosophy?" He suggested that the university should intervene now, before it was too late, as it was a well-established military practice to stop one's enemy before they had advanced too far "in soul or force," especially if they were "quick of tongue, of sharp intellect, subtle in their inventions, and desirous of glory." Implied here is the notion that Galileo's fame would attract students the way the Accademico Incognito had initially been drawn to the work. Unfortunately, however, they would be too naïve to realize that his fame was not actually based on merit but on attention seeking: Young people were thought to be more susceptible to novel, elegant ideas. As was often the case in debates between *novatores* and Aristotelian philosophers, the Accademico Incognito also worried about the effects on the university and its staff. He feared that attendance would dwindle and few would be left to listen to "great teachers who had Aristotle as their guide and first master."[45]

The four Pigeons may have hoped that the students themselves would pick up their treatises and, in this way, be prevented from straying from the right path. They also dedicated their works to four different members of the Medici family, most likely in the hope that its prominent members might intervene at the studio.[46] These efforts failed, however. A later document, from 1629, makes it clear that the studio hoped to profit from Galileo's appointment precisely because he was so well known and his fame would benefit the university. The document's author, Niccolò Cini, a Florentine canon, defended the fact that Galileo rarely lectured yet still had a high salary by pointing out that lecturers were not only appointed for the benefit of individual students. "One also searches," he wrote, "to have the most distinguished and famous professors for the reputation and honor of the university itself." Galileo had earned "advanced name and fame" through his publications, which in turn had been praised by "all of Europe's famous scholars, who have celebrated him with their writings, whose glory—no one can deny—reflects well on the Studio di Pisa, because Sr. Galileo is called its First Mathematician."[47]

This statement is of course not a response to Galileo's conflict with the Pigeon League, yet it reflects the way that conflict played out in the longer run and the effect it had on the different participants' reputations. While the dynamics of long-term fame prevent us from seeing the full picture—the sources at our disposal have mainly come down from Galileo and his supporters, his adversaries having left a much smaller paper trail—this imbalance is in itself significant. Galileo's four opponents did not succeed in discrediting him fully, and the sources that have come down to us further suggest that contemporaries found the Pigeon League's responses to Galileo to be too vicious and too driven by petty concerns. Cigoli dismissed Galileo's opponents as envious of his fame, and the moderate critic Tolomeo Nozzolini, who had lectured in logic at the Pisan studio before being appointed parish priest in Mugello, held a similar opinion. In a letter to Alessandro Marzimedici, the Florentine archbishop, he wrote that he did not agree with all of Galileo's points but also felt that "the league and the Anonymous proceed against him with tricks and do not wage a fair war."[48] The *lega*'s tactics, his letter suggests, had done more damage to their own reputation than to Galileo's.

The publishing history of Galileo's work even suggests that his conflict with the Pigeon League had a positive impact on his fame. The original dispute was undoubtedly local in character, and Galileo and his opponents all dedicated their works to different members of the Medici family, thereby signaling their regional loyalty.[49] Yet Galileo simultaneously attempted to transcend the original local setting of the dispute by sending out the work to scholars across Europe in the hope of winning wider recognition. He seems to have been successful. In the foreword to the second edition, the printer claims that the first print run had quickly sold out and he had ordered a second run in response to requests from interested readers in Venice, Rome, and other places. These remarks may have been part of a clever commercial strategy, but the fact remains that the work was reprinted within a year. None of Galileo's other works was reprinted within a comparable time frame, suggesting that this work's impact on Galileo's fame has been underestimated.[50]

Galileo thus largely survived the attacks of his four opponents unscathed. They were not the last ones he would have to deal with, however. In exactly this same period, a different (albeit related) group of opponents was preparing their own attack. This group of adversaries similarly criticized Galileo's fame, but their criticism went further and, as they had other media at their disposal, their offensive was more effective.

Pride and the Pulpit

Of the three Florentine Dominicans who led the offensive against Galileo around 1615, the one who most openly and frequently attacked Galileo was Raffaele delle Colombe, brother of the Pigeon League's leader, Ludovico delle Colombe.[51] He would speak against Galileo from his pulpit on several occasions around the same time that his academic opponents published their treatises. His objections to Galileo resemble those of the Pigeon League, but they were more severe and more dangerous: While he similarly accused Galileo of seeking individual, earthly glory, he did not stop at the allegation of ambition.

Delle Colombe's sermons exemplify the rich early modern media system and the porous boundaries between orality and print. The printed iterations of his sermons—sermons originally delivered orally—contain responses to Galileo's printed works, to a lecture Galileo delivered once before the Accademia Fiorentina, and perhaps also to the versions of his letters on sunspots that circulated in manuscript form in Florence. These sermons have come down to us in four large volumes, published between 1613 and 1627, and although these printed versions do not necessarily reflect their oral counterparts word for word, they are the closest we can get to the original sermons Delle Colombe delivered.[52] Delle Colombe most likely delivered them between 1609 and 1615, and the sermons all focus on the same topic: scholarship, the inappropriate desire for earthly glory, and pride.[53]

It was not uncommon for members of the clergy to discuss scholars and their work. For example, in 1604 a Jesuit in Padua attacked the teaching of Cesare Cremonini, Galileo's colleague at the studio, that suggested the mortality of the soul.[54] That scholars became the topic of sermons reflects the increasing public visibility of scholarship in this period, and it further contributed to scholars' renown or notoriety. The Dominicans were known for their preaching ability and had a strong position in Florence. Raffaele delle Colombe usually preached in the church of Santa Maria Novella, home of the Dominican order in Florence, but in 1615 he delivered a series of Advent sermons from the pulpit of the city's largest church, the Duomo.[55] In both churches he would address an audience whose size would far surpass the normal print run of the scholarly works written by Galileo or his opponents; the Duomo in particular may host many more individuals than the number of books in a normal print run in this time (which would rarely exceed a thousand copies). This, and the later publication of his sermons, meant that Delle Colombe's discussions of Galileo's recent sunspot discoveries reached an audience many times larger than the readership of the academic works and poems discussed so far. His audience would also be more diverse. Like other early modern preachers, Delle Colombe would tailor his sermons to resonate with these different types of listeners. For the learned and elite listeners who picked up on learned references and were already up to date on the latest news, preachers would include allusions to classical texts, the Bible, and current events to make the sermon's main themes more accessible, which intensified the audience's attention. At the same time, sermons also played an important role in transmitting news to the general public in attendance.[56] Delle Colombe's audience likely included people already familiar with Galileo, as well as people who had first encountered him through his sermons. He made sure to warn both against what he perceived to be Galileo's dangerous influence.

In the first sermon Delle Colombe focused on Copernican scholars and their improper desire for knowledge and glory. He delivered this sermon on the third Sunday of Advent, in or before 1609.[57] It thus preceded Galileo's telescopic discoveries as well as his move to Florence but indicates the backdrop against which his discoveries would be received. Delle Colombe opened that day by commenting on the day's snowy weather and its relation to the larger theme of his sermon, namely earthy glory: "Everything down here is full of the whitest snow, so of perceptible goods, and of earthly glory, of which snow is the symbol" (presumably because it melts away so quickly). The preacher then introduced the main topic of his sermon: the relation between curiosity, the wrongful desire for glory, and pride. He stated that being "too curious" in investigating heavenly glory only had a dimming, darkening effect, and he strengthened his point by quoting from the Book of Proverbs: *Qui scrutator est maiestatis opprimeretur a gloria*—"he who investigates greatness will be crushed by glory" or, in a more liberal translation, "for men to search their own glory is not glory."[58] Delle Colombe then continued to speak about the need for humility, as those who sought to achieve glory themselves came dangerously close to the sin of pride. This was the worst sin, Delle Colombe held, because all other sins always contained an element of pride, essentially in the refusal to obey God and the wish to shake off the "yoke of obedience."[59]

Delle Colombe chose to drive home this point by invoking the example of proud, ambitious scholars, who act like drunks inebriated not by strong wine but by undiluted knowledge. This made them act like fools, as he argued in a long passage interspersed with biblical references (in italics):

> Pride has clouded the vision in such a manner that one cannot understand even things that are nearby and very easy. . . . And what can be more sensible than seeing that *God established the world which shall not be moved*, and even so the Copernicans say that the earth moves and the heaven stands still, because the sun is the center of the world, and for this reason one can say that they suffer from vertigo, *The Lord has poured into their midst the spirit of confusion; and they have staggered as a drunken man staggereth in his vomit.* Even though science with the warmth of knowledge should be a reason to recognize God, nonetheless they, abusing it because of their pride, cool down, and give in to the fear, *They marveled, they were disturbed, fear took hold of them*, because they are afraid of losing earthly glory, not the heavenly, *they shook with fear where there was no reverence.*[60]

Highly critical of scholars whose claims ran counter to the Bible, Delle Colombe accused them of being blinded by pride. This, he said, made them pursue inappropriate questions and subjects and led them to defend claims that were absolutely ridiculous.

Not all scholars looking for knowledge succumbed to the sin of pride. Delle Colombe explained the difference between the proper and improper search for knowledge by pointing to Aristotle, who had held that it was magnanimous to "search for the truth of things" and "cared not for the praise of men." Thomas of Aquinas further distinguished between proper and improper glory: "Magnanimity is inclined to [spur one] to do honorable things, and worthy of honor, and is contrary to vainglory, which inordinately seeks glory."[61] Finally, Delle Colombe ended the first part of this sermon by citing several lines from Jeremiah 9:23 and 9:24. In these verses, he told his audience, the Lord says that the wise should not boast of their wisdom or the strength of their power. If one must boast (*gloriarsi*), let them boast that they know God and understand that he exercises kindness, justice, and righteousness on earth.[62] In some cases, then, the pursuit of glory was allowed and even praiseworthy. But Galileo, Delle Colombe's next sermons made clear, did not desire the glory that came from knowing God; he pursued the other, wrong kind of glory.

The Dominican started his open attack on Galileo with a relatively reticent accusation in a sermon delivered during a Lenten period sometime from 1611 to 1613.[63] The second part of this sermon focused on hell and its horrors and offered the perfect context in which to refer to a lecture Galileo gave more than twenty years earlier for the Accademia Fiorentina. In that lecture Galileo had focused on the mathematical dimensions of Dante's hell, and Delle Colombe now presented this type of inquiry as an example of the wrong sort of curiosity pursued by modern scholars: "I have always been astonished by these earnest measurers, especially by those who from the measured, as they say, proportion of the Spheres, and by multiplication and subtraction, argue that Hell is 7,865 miles wide." A note in the margins offered clarification for those who did not immediately understand the reference: "Galilei disputed the measure of Dante's hell in response to Antonio Manetti against il Vellutello."[64] The assumption that gauging the measurements of hell is important and therefore warrants scholarly attention was, to Delle Colombe, a perfect example of *superbia*. Measuring hell was a vain distraction from what is really important: not ending up in it.

From here on Delle Colombe's criticism of Galileo would only become more explicit, fueled by Galileo's growing fame and his continued contributions to astronomy. Delle Colombe especially objected to his work on sunspots. Galileo had communicated his observations and interpretations of the spots in three separate letters addressed to the Augsburg banker Marcus Welser over the course of 1612; he would publish his findings the subsequent year with the help of the Lincei. Eileen Reeves has pointed out that in their letters Galileo and his opponent Scheiner frequently drew on imagery related to hearing, the wisdom (or vanity) of crowds, and popular knowledge, with Galileo in particular clamoring for the support of the masses. Just before the book on sunspots came out, his close friend and supporter Ludovico Cigoli wrote to say that they should let the conflict "go all the way to the benches," as—rather improbably, Reeves adds—the Florentine masses would support them against Scheiner.[65] Delle Colombe's sermons suggest that any popular support Galileo tried to rally to his cause was met with fierce resistance.

In a sermon delivered on the first Sunday of Lent, possibly in

1613 or 1614, Delle Colombe portrayed Galileo's newest telescopic discoveries as a prime example of human pride in the light of divine wisdom. Note that he does not mention Galileo by name but seems to assume his audience knows to whom he is referring:

> Our inventive Florentine mathematician mocks all the ancients, who held the Sun to be absolutely pure and clear of even the smallest spot, thus giving rise to the proverb *Searching for the spot in the Sun*, but he, with the instrument he calls the Telescope, shows the sun to have regular spots, as he has demonstrated through observations of days and months. But this will be done more truthfully by God, because *The heavens are not pure in his sight*, if stains are found in the suns of the just, [what] do you think [might] be found in the moons of fickle sinners[?][66]

To the ancients, the idea that the heavenly bodies would not be pure was so ridiculous that they even had a proverb about it, as Delle Colombe said, and yet Galileo willingly and deliberately tried to prove that the heavens were stained.[67] The accusations here echo some of the claims of Galileo's academic opponents, who argued that he did not pay proper respect to the long tradition of scholarship that came before him. But Delle Colombe's allegations were more severe. Galileo was intent on undermining not just ancient knowledge but also Christian belief: Christian thought, based on Aristotelian philosophy, contrasted the perfection of the heavenly bodies (and God) to the imperfection of the lower terrestrial sphere (and humanity). The sun in particular was often read as a symbol of the godhead and the moon as an extension of the Virgin Mary.[68] Galileo, in his *Sidereus Nuncius*, had already called into question the perfectly smooth surface of the moon, and this was the aspect of his work that encountered the most enduring resistance from the Jesuit mathematicians at the Collegio Romano.[69] Now he also argued that the sun was *not* perfect but was in fact stained by spots, thereby further calling into question not only the division between the heavenly and the terrestrial but also the perfection of God and Creation.

Delle Colombe's biblical allusion to the notion that "the heavens are not pure in his sight" further added to the case against Galileo. At first sight it seems almost to condone Galileo's discoveries: If the heavens are not pure when compared to the perfection of God, perhaps there are indeed spots in the sun? Not so. The line is a reference to chapter 15 from the Book of Job, which stresses that God is so holy that even the heavens seem impure in comparison, that men know nothing of what God intended, and that those who claim to have knowledge of the world thereby "undermine piety and hinder devotion to God."[70] The Dominican ended his diatribe by reminding his audience that Galileo, like all men, would one day have to answer for his sins. If God finds spots even in that most perfect heavenly body, the sun, then sinners—and in particular those claiming to know the world based on mere days and months of observation—will not escape his judgment.[71]

The last two sermons further explore the precise threat that Galileo and sinners like him posed. These sermons again focused on his latest astronomical discoveries, and Delle Colombe delivered the sermons in December 1615 from the pulpit of the Santa Maria del Fiore church (the Duomo). This was a sensitive moment; as the next section shows, around this same time the Inquisition was nearing the end of its investigation into complaints against Galileo made by Tommaso Caccini and Niccolò Lorini, which had so far resulted in the decision to subject Galileo's recent work on the sunspots to an examination. In his sermon from the first Sunday of Advent, Delle Colombe spoke on Luke 21:25, which discusses how one would know the end of time was near: "There will be signs in the sun, moon and stars."[72] Delle Colombe seemed to think Judgment Day would soon arrive and urged his audience to make sure they had confessed their sins. He also told his listeners *why* the signs in the heavens had appeared:

> But do you wish to see the malice of sin, which even infects the heavens and makes the Sun black and obscures the Stars? St. Paul the Apostle said it, *Without the shedding of blood there is no forgiveness. It was necessary, then, for the copies of the heavenly things to be purified with these sacrifices, with better sacrifices.* Scripture is difficult here, and I with St. Thomas of Aquinas understand the Heavens in the actual sense, not the metaphorical. But how could the sinner stain the Heavens? disdaining it, turning his back [on it], almost as if the Heavens were mud.[73]

Delle Colombe thus explicitly portrayed the recent astronomical discoveries as sinful. Calling into question the perfection of the heavenly bodies through observation or measurement was a proud and vain endeavor, with grave consequences. The sinner's pride infected the heavens and thereby drew the end of time nearer, thus threatening the congregation's salvation. Delle Colombe could not but warn them; as preacher, it was his responsibility to ensure the congregation's spiritual health against threats such as Galileo.

Two days later Delle Colombe returned to the topic of the heavenly bodies and the appropriate way of observing them, this time in a sermon devoted to Mary. Citing Seneca as his source, he said that mankind had been able to observe the sun directly only from the moment mirrors were invented. He then drew a parallel between this mirror and Mary, who in a similar fashion had allowed mankind to see the beauty of her son.[74] Delle Colombe held that this was the only proper way of using an instrument: to observe the beauty and perfection of the celestial bodies (and by extension, God) without being blinded by them. The preacher then contrasted this proper way of looking at the heavens with another one: "Of someone who searches for a defect even where there is none, did the ancients not say, *He searches for the spot in the sun*? The sun is without stain, and the mother of the sun without stain."[75] The Dominican did not mention either Galileo or his telescope directly: the reference would be clear from context, especially as one of the words he used for mirror, *cristallo*, was frequently used for the telescope.[76] Once more, then, Delle Colombe accused Galileo of proudly and purposefully casting doubt on the perfection of the heavenly bodies, by his claiming that the moon's surface was rough and the sun spotted.

Like the academic opponents we have previously encountered, Delle Colombe felt the need to defend a specific community against the danger posed by Galileo's personal desire for glory. Yet his criticism of Galileo was even more biting and pronounced than theirs. The Pigeon League had denounced his personal ambition, and Ludovico delle Colombe at one point drew on religious imagery and undertones as he pointed out that those finding glory would "not just be admired but reputed to be Gods."[77] None of them, however, had gone so far as accusing Galileo of having succumbed to the sin of pride. Raffaele delle Colombe's attacks also formed a different sort of threat because he had a much larger platform than Galileo's academic opponents, in print but especially in his original, oral sermons. These, delivered in the Santa Maria Novella and in the Duomo, would also attract a wider subsection of society than the printed treatises Galileo's academic opponents used to attack him, and ironically, he may actually have enhanced Galileo's visibility in Florence even as he was opposing it. Whether Delle Colombe succeeded in turning his audience against Galileo is difficult to assess, but he spoke from an authoritative position and must at least have played an important role in mediating the views of people with no prior knowledge of Galileo and his work. Finally, Delle Colombe's sermons were also particularly dangerous to Galileo as he could not easily defend himself against attacks from the pulpit. While he could potentially address a relatively large audience from his lectern at the Pisan university, he needed to lecture only on special occasions and spent most of his time on research and writing. The students he could reach, moreover, made up a much smaller section of society than Delle Colombe's audience.

As Galileo became more well known and was discussed beyond the original, academic context in which he had won his fame, he also lost the ability to reply to some of his newer, more powerful and dangerous opponents. In some settings, then, the dynamics of fame worked against him, and Delle Colombe's attacks posed a larger risk to Galileo than the attacks of the Pigeon League. This risk only became more pronounced in the second line of attack Galileo's Dominican opponents followed.

False Claims and False Fames

All three of Galileo's most pronounced Dominican opponents—Raffaele delle Colombe, Niccolò Lorini, and Tommaso Caccini—publicly and loudly preached against him from their pulpits in Florence. At the same time, Lorini and Caccini also sought to counter Galileo in a more private and ultimately more dangerous manner: In the early months of 1615, they denounced Galileo to the Inquisition in Rome. Their most important allegation against him was that he was increasingly known as the leader of a growing movement in Florence that sowed untruths and possibly heresies in the streets of the town. This assault was not only more formal but also articulated behind closed doors, and this combination was precisely what made it so dangerous to Galileo: it left him in the dark as to the best way to defend himself. The men were not particularly successful until Galileo, in his attempts to counter their complaints, unwittingly confirmed precisely the image they had sketched of him. This complex interplay between rumor, *publica fama*, Galileo's growing fame, and his own actions was one of the reasons leading to the 1616 ban on Copernicus's work and the injunction delivered to Galileo ordering him not to discuss Copernicanism.

Lorini approached the Roman Inquisition in February 1615. He was concerned about a piece of writing that circulated widely in Florence and that he thought contradicted Scripture. Earlier, in 1613, Galileo's student Benedetto Castelli had participated in a debate at the Medici court revolving around Galileo's recent discoveries and their relation to Scripture. Castelli of course claimed that Galileo's discoveries were in line with the Bible and its teachings; the Grand Duchess Christina, whose curiosity had sparked the debate, argued against him. Like the discussion at Salviati's, this one did not remain confined to its original setting either, and we can again follow the debate's development from an oral discussion to a written one. Galileo, after hearing from Castelli about the debate, chose to weigh in via a letter, dated 21 December 1613, in which he listed all his arguments to prove that his findings did not contradict the Bible. The letter was never published, but it circulated widely in Florence nonetheless. While it was addressed to Castelli, early modern letters were rarely private documents in the modern sense of the word. Galileo was well aware that his letter was read by people other than Castelli and probably even intended it; he later took care to ensure its circulation in Rome as well.[78]

To his opponents, however, the letter's uncontrolled circulation was highly problematic. Among those who obtained a copy was Lorini, who in his complaint to the Inquisition emphasized both the letter's controversial content and its wide circulation. It was clear to him and "all the Fathers of this most religious convent of San Marco" that the letter contained passages that appeared suspicious or rash. Yet, as no one had taken action, the letter was to be found "in the hands of all."[79] Now that Galileo's ideas were allowed to spread without opposition, his following was growing: "I have to say that I think all of them, who call themselves Galileisti, are good men and good Christians, but a bit presumptuous and harsh in their opinions." Like Delle Colombe, Lorini thought Galileo to be a threat to the maintenance of religious orthodoxy in the city. The problem was that the Galileisti did not keep their wrongful opinions to themselves but were going around in the streets "saying and sowing thousands of impertinent things, only to seem witty."[80]

The Inquisition quickly set to work after receiving Lorini's letter. They initially limited their investigation to establishing the possible heresies contained in Galileo's letter to Castelli, and after examining the copy sent to them by Lorini they declared that nothing other than three relatively minor items in it contradicted Scripture.[81] Still, they did order the Pisan inquisitor to confiscate Galileo's original letter. Castelli no longer had it, however, and the Pisan archbishop and inquisitor informed Cardinal Mellini in Rome that they would try to retrieve the letter from Galileo. Castelli wrote Galileo several times on this subject, and from this moment onward both men became much more careful in handling their correspondence, having been alerted to the potential risks of circulating semipersonal letters. Galileo did not comply with the request to share his original letter but instead sent to Rome a toned-down version of his letter to Castelli.[82]

At this point the Inquisition might have dropped the investigation altogether, if it were not for Caccini's deposition. Caccini arrived in Rome in February and appeared before the Inquisition on 20 March, where he confirmed Lorini's depiction of Galileo as a growing threat to religious orthodoxy in Florence.[83] In his opening statement, Caccini stressed that in Florence it was *publicissima fama*, or very publicly known, that Galileo—unchecked by any authority—held and taught the heliocentric worldview. The Inquisition's subsequent interrogation focused on assessing the gravity of the accusations and establishing the credibility of Caccini's deposition by way of checking his sources.[84] Caccini's answers revealed that he had neither met Galileo nor heard him speak, that he did not know whether Galileo taught publicly or had many students, and that he had obtained most of his information from others. He was most precise when asked how he knew Galileo was "teaching and holding" possibly dangerous opinions: He referred to "publica fama" and named two additional sources—the bishop of Cortona and a student of Galileo, who was "a certain Florentine gentleman degl'Attavanti." This student had been heard defending three heretical opinions by a Fra Ferdinando Ximenes, regent of Santa Maria Novella.[85]

The allegation that Galileo was teaching heretical opinions was especially severe; this was a much graver offense than simply holding such views. Caccini confirmed Lorini's assessment that Galileo was raising a following in Florence, one that was growing more and more brazen: He claimed they had even gone so far as to approach the preacher of the Duomo and ask him to publicly rebuke Caccini for delivering his anti-Galilean sermon on Joshua's request that God stop the sun's movement.[86] If this was not enough cause for concern, Caccini also casually mentioned ties between Galileo and Fra Paolo Sarpi, the Venetian troublemaker whom the pope and the Inquisition had tried to silence a few years earlier via an attempted assassination. The inquisitors asked Caccini if Galileo was a good Catholic, and Caccini answered that many thought him a good Catholic, while others suspected him in matters of faith because "they say he is close to that Servite Fra Paolo, [who is] very famous in Venice for his impieties, and they say that even now they exchange letters between them."[87]

Lorini's letter and Caccini's deposition point to concerns about the dangers of talk in public spaces like streets and squares. These were spaces, as historians have pointed out, "where different groups strove to have their voices heard" and where social and institutional order were articulated, asserted, and contested.[88] Talk and rumor could offer marginalized groups a way to express their discontent and undermine authority, and established authorities consequently feared the popular aggression and riots they might spark. This spurred the English philosopher and statesman Francis Bacon to write that "rebels . . . and seditious fames and libels, are but brothers and sisters."[89] He doubled down on that sentiment in an essay on seditions and troubles: "Libels and licentious discourses against the state, when they are frequent and open; and in like sort, false news often running up and down to the disadvantage of the state, and hastily embraced; are amongst the signs of troubles."[90] To authorities aiming to maintain political or religious order, the dangers of talk were enhanced by its ephemeral quality: this made it difficult to trace and contain voices of discontent.[91]

The cumulative effect of loose tongues only made matters worse. Talk previously shared among just a select few could easily become publica fama, or knowledge shared within a community. As we have seen, this could serve as evidence or as grounds for investigation in court cases and was thus linked to practices of truth finding. However, Galileo's opponents were especially concerned that such talk would reach people who could not discern between truth and falsehood and who would take the increased presence of Galileo, his students, and his claims as signs of their truth—a situation that would threaten the religious orthodoxy in the city. Lorini's and Caccini's complaints, then, were appeals to the appropriate religious authorities to check the speech circulating in Florence, renegotiate what could and could not be said in public, and assert order.

Now that the Inquisition had two sources that portrayed Florence as a hotbed of insurgency, with Galileo as its focal point, their next step was to verify these claims, and they especially concentrated on finding students who had been taught by Galileo and could confirm the Dominicans' story. Thus, the Roman office asked its Florentine counterpart to question several people Caccini had mentioned: Father Ferdinando Ximenes, a Dominican who had told Caccini about the Galileisti's contact with the preacher of the Duomo, and the student Gioanozzo Attavante, who had apparently let slip that Galileo's teaching went against Scripture. The Florentine inquisitor did as he was told, even though he did not seem to find the matter particularly urgent. In April he let his Roman colleagues know that some of the witnesses "were busy with the Lent sermons," and in May he wrote that the investigation was further delayed because Father Ximenes was out of town. He was the key witness, the Florentine inquisitor thought, as he could tell them about "the three propositions which are claimed to be uttered by the students of said Galileo, which is the main foundation of what can be held against said Galileo and only needs proof."[92] The Roman Inquisition office then wrote to Milan to see if the inquisitors there could get hold of Ximenes. By the time their letter reached the Milanese office, however, Ximenes was back in Florence, where he was finally questioned on 13 November 1615.[93]

Father Ximenes did not confirm but rather allayed the suspicions raised by his colleagues, the Dominican friars. He told the

Inquisition that while he had never met Galileo, he had heard about Galileo's opinions and indeed thought that they seemed to contradict "true theology and philosophy." He had even heard some of Galileo's pupils say that "the earth moves and that the heavens are immobile." However, he added that he was not sure whether they were giving their own opinions or echoing those of Galileo. Ximenes could also only give the Inquisition the name of a single student, Attavante, whom Caccini had already named. Ximenes said that he and Attavante had discussed Galileo's discoveries several times in his room at the convent of Santa Maria Novella, occasionally in the presence of others. Yet Ximenes believed that Attavante had only said "the above-mentioned things" for the sake of argument (*disputationis gratia*). He thought the student only possessed superficial intelligence of the matter and deemed it likely that Attavante was repeating Galileo's opinions and had not himself come up with the ideas he expressed. Even so, Ximenes said that he had immediately reproached Attavante for these false opinions. This seemed to satisfy the inquisitor.[94]

Attavante's hearing, held the next day, further lessened suspicion toward Galileo. He told the Inquisition that he had never actually studied under Galileo and that they had only exchanged letters. He also said he had never heard Galileo say anything against Scripture, although he had heard him defend the Copernican worldview. The Inquisition pushed him on this: Had Attavante ever heard Galileo interpret Scripture according to his own convictions, and had he ever defended one of the propositions that had been included in Caccini's deposition? Attavante answered that he had indeed heard Galileo give an alternative interpretation of the Book of Joshua (the Inquisition did not have further inquiries on this matter) and that he himself had defended several of the given propositions—but only *per modum disputationis.* He added that Caccini, whose room was next to that of Ximenes, had perhaps overheard him and assumed that he was repeating Galileo's opinions, which was not true. Finally, the Inquisition inquired into Attavante's opinion on Galileo's Catholicism and asked whether he felt any enmity toward Caccini. With Attavante's answer to the last question, the hearing came to an end, and the Florentine inquisitor sent transcripts of these testimonies to his Roman colleagues.[95]

At this point, the investigation seems to have lost urgency. The accounts of Ximenes and Attavante did not corroborate Lorini's and Caccini's statements, and the Inquisition's concern about Galileo's growing following subsided. Their only follow-up in this investigation was to order an examination of Galileo's sunspot letters.[96] However, just three months later, in February 1616, Galileo was served with a special injunction that prohibited him from teaching or discussing heliocentrism except hypothetically. In addition, more than seventy years after its publication, Copernicus's *De Revolutionibus* was put on the Index of Prohibited Books.[97] One reason for this new phase in the Church's relation to heliocentrism was the way Galileo's behavior, and especially his high visibility in Rome, confirmed the rumors about him that Caccini and Lorini had relayed to the Inquisition.

Over the course of 1615, Galileo had remained in touch with friends in Rome so he could stay abreast of the latest developments in the Inquisition's proceedings. Word of Lorini's letter and Caccini's deposition had traveled fast, and Galileo was worried their allegations might have severe implications: He could be accused of heresy or be ordered to stay silent on certain subjects. However, rather than maintaining a low profile, Galileo responded by sending his work to Rome and asking a friend, Piero Dini, to circulate it there. This happened shortly after the Roman Inquisition had tried to collect Galileo's original letter to Castelli from its recipient, in February 1615. Galileo especially hoped that the Jesuits were open to his arguments—likely because of their earlier support of his work but perhaps also because of their long-standing rivalry with the Dominicans—and he asked Dini to make sure that they received a copy.[98] Just as the Roman Inquisition had been alerted that a piece of controversial writing by Galileo circulated freely in Florence, he then took care to ensure it was also widely read in Rome.

Galileo did not stop there. When he wrote to Dini about his letter to Castelli in February, he also told his Roman friend that he was preparing another piece of writing. This letter (which has come to be known as the Letter to the Grand Duchess) would show, through biblical interpretation, that the Copernican theory did not necessarily contradict Scripture.[99] Wanting to know how such a piece of writing would be received by the Church, Galileo asked to contact Rome's most important theologian, Cardinal Roberto Bellarmine, who was a Jesuit, and find out his opinion on the Copernican theory.[100] Dini did as he was asked, and in his response he encouraged Galileo to follow Bellarmine's explicit invitation to share such a piece of writing with him privately, before sending it to others.[101] Bellarmine also communicated a more implicit warning to Galileo, by way of a letter to Paolo Antonio Foscarini, a Neapolitan Carmelite. Foscarini had published a book in favor of Copernicus's doctrine in February 1615 and had sent a copy of it to Cardinal Bellarmine, hoping that he might help the Copernican cause.[102] Bellarmine, however, responded by way of a warning disguised as a compliment. It was directed to both Foscarini and Galileo: "First, I say that it seems to me that Your Paternity and Mr. Galileo are proceeding prudently by limiting yourselves to speaking hypothetically and not absolutely, as I have always believed that Copernicus spoke."[103] While stressing that it was best to discuss the Copernican theory only hypothetically, Bellarmine did imply that it might eventually be treated in a nonhypothetical way. For this, the Church would first need sufficient proof, which would need to be treated carefully to make sure Scripture could be reinterpreted correctly. Cardinal Bellarmine added that he did not believe sufficient proof existed, and he would not believe it "as long as it is not shown to me."[104]

Historians disagree over the meaning of this addition and the letter more generally; some argue that Bellarmine firmly rejected the possibility that the Copernicans could soon present him with sufficient proof, while others interpret Bellarmine's words as allowing for the possibility that proof might be found.[105] But whatever Bellarmine's intention, Galileo took this passage, which he likely received in April from Dini, to be an invitation.[106] He replied in a highly unsuitable manner, with his now finished letter to the Grand Duchess Christina. This did not give Bellarmine the proof for which he had asked, and instead Galileo argued that theologians should rely on the expert opinion of mathematicians when interpreting Scripture.[107] He also did not first send it to Bellarmine. While the Roman Inquisition was investigating accusations that Galileo had let a heretical text circulate freely in Florence, then, Galileo not only had this same text (although likely in a tone-down version) circulating in Rome but also produced another text that expounded on the arguments delivered in the first text.[108]

As if this were not enough, Galileo then traveled to Rome himself. He decided to do so at the end of November 1615, as he became aware that the testimonies of Ximenes and Attavante had just been sent to the Roman Inquisition. While we know that these accounts took the sting out of Lorini's and Caccini's accusations, Galileo was not aware of the contents of their testimonies; they were, after all, delivered in private. He therefore feared that the Dominicans' attempts to discredit him with the Inquisition were gaining rather than losing ground. Ignoring the advice given to him by the Florentine ambassador, Piero Guiccardini, Galileo arrived in Rome in December 1615.[109] Shortly after his arrival he wrote yet another work, the *Discorso del flusso e riflusso del mare*. This again may have been an attempt to provide Bellarmine with the sort of proof for which he had asked, but Galileo again did not first send it to the cardinal. Instead, he asked the young Cardinal Alessandro Orsini to present it directly to Pope Paul V. This was a misjudgment. The pope was notoriously both conservative and ill-tempered, and Galileo had explicitly been warned by the Florentine ambassador that Rome was not "the place to dispute on the moon" nor "to support or bring new theories to."[110]

But that was indeed precisely what Galileo had come to Rome to do. In early February, not long before he had Orsini present his work to Paul V, Galileo wrote home to say that he had already achieved his goal and saved his reputation. In that same letter, however, he wrote that he wanted more and that he saw himself as uniquely suited to establishing the truth of Copernicus's theory once and for all. For eighty years, he told the Tuscan secretary, a whole community (*università*) had been trying to defend this "doctrine and opinion" through "printed works, private writings, public reasonings and preaching, and in private conversations." As the "science taught by me" confirmed the truth that needed to be acknowledged, he wrote, it was his obligation to help them get the truth recognized. Galileo decided to stay and advance the new issue that, as he put it, had "been linked to my cause."[111] Unfortunately, presenting the work through Orsini confirmed once more that Galileo was actively seeking to enlist others to aid in his cause. The Florentine ambassador included a damaging report on the meeting with Paul V in his letter to the Tuscan secretary, lamenting that Galileo ignored all the advice about "remaining quiet, without trying so hard to convince others to believe the same" when it came to his Copernican beliefs.[112]

Just days after Orsini's meeting with the pope, Galileo was called on 26 February to have a private meeting with Bellarmine. The day before, the cardinal had been ordered to tell Galileo to abandon his opinions and to "abstain completely from teaching or defending this doctrine and opinion or from discussing it."[113] Bellarmine spoke with Galileo the day after and told him, in the presence of several witnesses, "to abandon completely the above-mentioned opinion that the sun stands still at the center of the world and the earth moves, and henceforth not to hold, teach, or defend it in any way whatever, either orally or in writing."[114] In early 1616, Galileo was doing precisely what Caccini and Lorini said he had done a year earlier: operated as the leader of a growing movement, apparently eager to step up and rally new supporters to his cause. Different forms of fama had become entangled: talk, rumor, and publica fama about Galileo as the leader of a growing movement were confirmed by Galileo's high visibility. The way Galileo handled this specific attack on him shows the risks and limitations of his fame when confronted with allegations uttered behind closed doors: Not knowing what precisely he stood accused of, he could not defend himself effectively, and his actions only confirmed suspicions that he was trying to spread the Copernican theory. Having started in Florence, he now seemed to be trying the same thing in Rome, while the earlier publication of Foscarini's work indicated that his ideas were gaining a following in Naples as well.

In March 1616, Galileo told the Florentine secretary that Foscarini's book had been put on the index and that Copernicus's work would be corrected. A few days later he also told him he had met with the pope, who had assured him that he was not in any danger and that the papal court did not listen to slanderers.[115] The Florentine secretary wisely replied that the grand duke and grand duchess had been glad to hear of his happy meeting with the pope, that it seemed to them that "his reputation was now intact," and that they therefore asked him "to quiet down and not speak of these matters any more, and instead return home." This was all for his own good, he added: "They say this for your benefit and your peace."[116] Galileo lingered in Rome a little longer, before being firmly recalled to Florence in May.[117] In the meantime, however, another of fama's fickle twists of fortune meant that his meeting with Bellarmine had given rise to rumors that he had been forced to abjure. To counter the murmurs, which had made it all the way to Venice and Pisa, Galileo finally needed to ask Bellarmine to issue a written statement to counter these speculations. This statement, and Bellarmine's earlier warning, would later come to play a role in the 1633 trial.[118]

In tracing the responses of two sets of Galileo's adversaries—academics and Dominicans—this chapter has revealed that Galileo's fame evoked fierce opposition from groups that interpreted his growing visibility not as a sign of merit but as an indication of his moral shortcomings and the danger he posed to established academic or religious authorities. These charges were interrelated; by drawing on fame's associations with vanity and pride, his opponents cast Galileo as unreliable and determined to do whatever it took—including undermining long-established truths—to garner attention.

The dynamic between Galileo and his audiences further alarmed his opponents. The high visibility of his work, they feared, meant that it easily reached people who were curious to learn new things but unable to distinguish truths from falsehoods and who wrongly mistook Galileo's fame as a sign of quality. These ways of thinking about crowds as dangerous and fame as a sign of pride lost much of their poignancy over time, but they precede later ideas about the public as driven by a desire for superficial gossip. While this concern for the public's gullibility led his opponents to position themselves as protectors of their respective communities, their attitude toward the public also reveals fears about its potential power to undermine order and orthodoxy. The resulting suspicion of uncontrolled circulation of information and ideas extended not just to the printed works that tend to be associated with fame but also to manuscript letters and oral discussions. The boundaries between these different forms of communication were porous at best, and to trace discussions about Galileo's work and the value of his fame, we need to look beyond the world of printed texts to that of the pulpit, the public squares, and the streets. This approach shows that Dominican friars especially objected to the open, public way in which Galileo's claims about the heavenly bodies—which they felt might amount to heresy—were discussed. To counter this unchecked circulation of dangerous ideas, they not only lodged formal complaints about Galileo but also spoke up in public. Though this meant putting the spotlight on these issues once more, it was also a way to reassert their authority over matters of Scripture and to negotiate the limits of what could be said publicly and by whom. Ultimately, debates over the proper kind and forms of fame were debates over whose judgment would count when it came to deciding which contributions were worth celebrating. The next chapter shows us how these debates played out in the aftermath of the trial.

5

GLORY OR INFAMY

ON 13 AUGUST 1643, EVANGELISTA TORRICELLI SURPRISED HIS audience at the Accademia della Crusca by saying that posthumous glory was worthless. In fact, he advised the learned members of his audience to reap fame's benefits during their lifetime, because "fame is alive to the living, and dead to the dead."[1]

Torricelli, who had moved to Florence to meet Galileo and lived with him in his final months, posited these unusual and controversial claims in a lecture entitled *Della Fama*, delivered a year and a half after Galileo's death. The Roman mathematician had been present during the modest, subdued burial in the Santa Croce church in Florence, where Galileo was not buried in the family grave but in a smaller room on the right side of the church.[2] Torricelli was then named Galileo's successor as grand-ducal mathematician and appointed a member of the Accademia della Crusca. In that capacity he delivered several lectures, most of which dealt with mathematical subjects.[3] His lecture from August 1643 stands out not only for its subject but also for the unconventional view on fame Torricelli shared with his audience. It was a widely accepted commonplace that the only type of fame worth striving for was posthumous; earthly fame was easily dismissed as fickle, fleeting, and fit only for the vainglorious, while posthumous fame, or glory, counted as the just reward for extraordinary achievements. Why then would Torricelli proclaim the opposite, going so far as to tell his fellow academicians that "after the final funerary rites, all men will become equally famous"?[4]

A possible answer is that Torricelli was only joking. The opening of his lecture referenced Carnival, and he explicitly mentioned that his lecture would appear to be a "satirical invective." His audience seemed to agree; the academy's secretary later noted in the group's minutes that Torricelli had delivered a "paradoxical lecture," or *lezione paradossica*.[5] Yet, this explanation is only partly satisfactory. While the lecture seemed jocular in tone, it belied a deep and heartfelt frustration on Torricelli's part about the way Galileo was remembered in Florence. Only a year and a half after his death, Torricelli lamented, his teacher had all but passed into obscurity. With his carefully timed lecture, he now hoped to spur his fellow academicians into righting this wrong. But had Galileo's contemporaries really forgotten him?

They had not. As it turns out, Torricelli's lecture was one of many creative attempts by Galileo's supporters to resist and reshape the narrative the Roman Catholic Church drafted with the 1633 trial. Galileo's *Dialogo* and the subsequent trial, in which he was found guilty of "vehement suspicion of heresy," are the subject of hundreds of scholarly books and articles. Historians have offered widely diverging interpretations on a number of issues, including the severity of Galileo's sentence in relation to his transgressions, the link between the events of 1615–1616 and the trial of 1633, the role the Dominican and Jesuit orders played in Galileo's conviction, and the positions of specific members of the Inquisitorial committee, Francesco Barberini in particular.[6] It is impossible to revisit all these debates here or to give a full account of the trial and all the events leading up to it. Instead, the first section briefly covers the main events leading up to Galileo's conviction, after which the second section, building on recent works that have approached the trial from the perspective of legal studies, situates the trial and sentence in the context of infamy or memory punishments.[7] Such punishments seek to redirect the way convicts are seen and remembered by both their contemporaries and future generations, which is precisely what the Catholic Church aimed to achieve with four different measures it took during and after the trial.[8] Homing in on these punishments allows us to explore the role of authoritative institutions in shaping the long-term memory of individuals and to investigate the dynamics between legal practices and cultural traditions. The final part of this chapter shows that the Church's attempts to reshape the way Galileo was remembered did not go unchallenged. The Catholic Church was not the only party invested in Galileo's memory, and his supporters countered the efforts to cast Galileo in a negative light as best they could. This led to a protracted struggle, fought in private and in public, over the way Galileo should be remembered: as an infamous penitent or a scholar deserving of glory.

THE TRIAL

Galileo published his *Dialogo sopra i due massimi sistemi del mondo* (*Dialogue on the Two World Systems*) in 1632. Initially relieved to have finally finished the book he had been thinking of for a decade or so, Galileo would soon realize that the obstacles he had had to overcome during the long publishing process were nothing compared to what came next. Only a few months after the book's appearance, sales were suspended and the Florentine inquisitor summoned Galileo to his office, where he was told his colleagues in Rome had sent for him. Galileo managed to delay his departure by a few months, due to medical issues and the ongoing plague epidemic, but he eventually arrived in Rome in February 1633. Just five months later, on 22 June 1633, the Holy Office of the Inquisition found him guilty of "vehement suspicion of heresy."[9]

In the *Dialogo*, Galileo discussed the merits of the Copernican and Ptolemaic worldviews. The book took the form of a dialogue among three characters—Sagredo, Salviati, and Simplicio—who at first glance appeared to simply exchange arguments without reaching a clear conclusion. Such a treatment should theoretically have been possible. In February 1616, an Inquisitorial committee had come to the conclusion that heliocentrism was "philosophically and scientifically untenable and theologically heretical."[10] However, no decree conveying this decision to the public was published at the time. Instead, a month later a different institution, the Congregation of the Index, published a decree with a slightly different content. It did not use the word *heresy* but stated that the doctrine that the earth moved and the sun was without local motion was not in line with Scripture and could therefore not be held.[11] This decree also stipulated that Copernicus's book was suspended until it was corrected and cleansed of passages that described the earth's motion as a theory. Otherwise, the work had been left mostly intact, as it was quite useful as a calculating device.[12] The decree, as Maurice Finocchiaro has pointed out, left some things unsaid, in particular whether Copernicanism could be discussed only if one did not consider it a doctrine.[13]

This question was foremost in Galileo's mind as he went to Rome and met several times with the newly elected Pope Urban VIII in 1624.[14] Even though they were old acquaintances, Galileo did not broach the subject directly with the pope but instead relied on a broker, Cardinal Eitel von Hohenzollern. The German cardinal, Galileo told one of his correspondents, had heard the pope say that heliocentrism had not been declared a heresy, nor would it be anytime soon, but was only deemed "temerarious."[15] Galileo was cautiously optimistic and set to work, believing he could write his book as long as he made it clear he adhered to the Ptolemaic, geocentric worldview. He again went to Rome in 1630 to obtain the necessary permissions to have his work printed there. After showing the manuscript to Father Niccolò Riccardi, who was Master of the Sacred Place and the man who had also approved the printing of *Il Saggiatore*, Galileo was given provisional permission but instructed to change a few things. Galileo returned to Florence and prepared to resubmit the manuscript for final approval in Rome. However, when the outbreak of the plague complicated shipping between Florence and Rome, Galileo instead lobbied to have the work checked and printed in Florence. Riccardi eventually agreed to his proposal and sent instructions to the Florentine inquisitor, who approved publication, and in February 1632 about a thousand copies of the book were printed. The book was quickly distributed to various Italian towns, but continuing measures against the plague meant it took until the end of May before Galileo could send the work to Rome. Before that, in April, he had already sent a copy to his friend Elia Diodati in Paris, and he hoped to send more copies to Lyon.[16]

However, it quickly became clear that the dialogue left little doubt as to Galileo's true opinions on the topic of heliocentrism. Simplicio, the character who defended the Ptolemaic, geocentric worldview, was by far the dumbest of the three, and Sagredo's and Salviati's arguments in favor of heliocentrism were simply too convincing. Before long, the Holy Office had two separate reports drawn up, both of which posited that Galileo's work definitely amounted to heresy, as it did not discuss Copernicanism in a way that could be considered hypothetical.[17] In late September, Galileo was called to Rome, while Cardinal Francesco Barberini wrote a worried letter to the Florentine nuncio, asking whether Galileo had sent more copies of the book abroad.[18] This was a highly sensitive moment for the Barberini family and especially for Pope Urban VIII, who—amid the political turmoil of the Thirty Years' War—was accused of being too lenient toward heretics. The concern over the precise circulation of this work by an already famous, highly visible, and controversial author confirms that the pope and his nephew were worried that they might seem weak, so they felt a need to deal with Galileo in a way that reasserted their power.[19]

During the initial investigation, the inquisitors found further damaging evidence against Galileo in their own archives: the injunction Galileo had received in 1616. The Inquisition minutes showed that Cardinal Bellarmine had been tasked with personally ordering Galileo "to abstain completely from teaching or defending this doctrine and opinion or from discussing it." This new discovery suggested that Galileo had knowingly violated a personal warning addressed to him, aware that not heeding the injunction would lead to his imprisonment.[20] During the first deposition of the trial in April 1633, Galileo confirmed that he had been given an injunction in 1616, but he also presented his interrogators with a written document by Bellarmine: the letter, sent by the cardinal at Galileo's request shortly after their meeting, which only said that Galileo could not defend or embrace the Copernican doctrine.[21] This letter differed from the Inquisitorial minutes as it did not explicitly forbid Galileo to *discuss* heliocentrism in any way. Galileo claimed that Bellarmine's written communication had replaced his memory of the meeting, which had, after all, taken place more than fifteen years earlier. He also maintained that he treated the heliocentric model in a hypothetical manner rather than defend it as a convincing theory.[22]

Not long after the first deposition, the man in charge of Galileo's trial, Vincenzo Maculano da Firenzuola—Commissary General of the Holy Office—wrote to the pope's nephew, Francesco Barberini, to update him on recent events. His letter shows that the inquisitors carefully considered the best course of action given the circumstances: If Galileo did not confess to his wrongdoing, the papal official wrote, the matter could not be dealt with "as discussed." He had therefore taken it upon himself to deal with the matter "extrajudicially" and had conveyed to Galileo that he would need to confess to having treated heliocentrism as more than a hypothesis. According to the Commissary General, their talk had helped Galileo see the error of his ways; he was ready to confess and had only asked for a bit of time to carefully consider how best to make his "honest confession."[23] The papal official was relieved, as the Holy Office needed to strike a careful balance between seeming too lenient and too harsh. The general commissioner was especially aware that they needed to take Galileo's old age and reputation, as well as the position of the grand duke, into account. If Galileo confessed, the Holy Office could follow a suggestion made by Cardinal Barberini: "In this manner the case is brought to such a point that it may be settled without difficulty. The Tribunal will maintain its reputation; the culprit can be treated with benignity. . . . With this done, he could be granted imprisonment in his own house, as Your Eminence mentioned."[24] A few days later Galileo indeed changed his story. According to his second deposition, he now realized that he had not treated the Copernican model hypothetically. His "vain ambition" had gotten the better of him, he said, and he cited Cicero when he admitted that "I have more than my proper share of desire for glory." This had led him to proudly seek to display his own wit by making the arguments in favor of the heliocentric model as cleverly as possible.[25]

Galileo's confession of his own wrongdoing was crucial for the way he would be punished. After abjuring and cursing his false beliefs on 22 June "with a sincere heart and unfeigned faith" and swearing to never hold them again, he avoided incurring the heaviest punishments.[26] His abjuration enabled his conviction for "vehement suspicion" of heresy rather than formal heresy. In the official hierarchy of heresies this was a lighter offense, and it meant that he would not be condemned to full forgetting and infamy.[27] However, the way he was remembered would still need to undergo significant changes.

Memory and Infamy

In canon law and Christian thought, heresy was conceptualized as a form of treason. The heretic had not only betrayed God but also the *respublica Christiana* as a whole.[28] The common punishment for this specific form of treason was infamy, as the widely read Inquisition manual by Eliseo Masini, a Dominican theologian, tells us: "It is of such an ugly, and of such horrible character, the crime of heresy, that who commits it, incurs *infamia iuris, & facti*."[29] These two concepts, infamia iuris (infamy of the law) and infamia facti (infamy of the fact or act), were both rooted in Roman law and over time had been absorbed into canon law.[30] They reflect the dual meaning of the word *infamia* as both a formal legal standing and a more informal social status. Infamia iuris was administered by a judge and resulted in the loss of legal capabilities (*fama legalis*); it stripped the convicted of the right to leave a will and to act as a notary. People who had incurred infamia iuris could also no longer act as a witness. Infamia facti on the other hand was a socially imposed judgment with the broad aim of severely diminishing a person's social standing.[31] Through infamia facti, the community shared in the responsibility of safeguarding society from individuals who threatened its values. Seldom do punishments so perfectly fit the crime: Heretics had turned their back on their fellow Christians by committing heresy, and as a result precisely that community—thereby maintaining its own spiritual health—would exclude them from its ranks.[32]

But heretics did not lose only their fama. The Catholic Church also targeted their *memoria*—the way they were remembered posthumously: "Damn the memory of the dead heretic; even though, while alive, he has not been defamed by heresy."[33] Masini's manual provides an example of a sentence that might be imposed in such cases: "We declare him infamous, and excommunicated, and unworthy of an Ecclesiastical burial: and therefore we order his bones, if they can be distinguished from the bones of the faithful, to be unearthed, and brought outside of the Cemetery, and publicly burned in detestation of his grave crime. In addition, we release to the secular arm the effigy of the aforesaid N. [name] here present, so that it will likewise (as is appropriate) be burned."[34] Of course, if heretics were still alive at the time of their conviction, trying to separate their bones from those of the faithful was unnecessary; in those cases they could simply be denied Christian burials and resting places. In the most severe cases they were burned alive and their ashes cast into a river so that no trace of their bodies would ever be found on earth.[35]

These measures taken to mediate a person's remembrance possess clear similarities with sanctions used in ancient Roman law. They especially overlap the practice of *damnatio memoriae*: a memory sanction applied to members of the elite who had been convicted of political treason.[36] Under this practice, all signs of public glory and recognition were destroyed in order to mark the convicted person's loss of honor before the community. Sanctions could take different forms, but common effects included exile; the destruction of representations like busts, portraits, and statues of the convicted transgressor; and the erasure of names from monuments and documents. A prohibition on the performance of standard funerary rites and public mourning for the disgraced individual meant that memory sanctions affected not only the posthumous glory of those convicted but also the status of their families, who were marked as secondary members of society through their exclusion from important social rituals.[37] Most often, complete forgetting was not the goal; instead, the sanctions aimed to establish a marked forgetting or the redirection of memory. As public signs of honor came down, traces of the original monuments remained visible and served as reminders of the convicted person's punishment and subsequent loss of honor. They thereby became monuments once more, only now they sought to deter rather than to inspire, communicating to the wider community that certain kinds of behavior would not be tolerated.[38]

This sort of marked forgetting and marked remembrance was also the purpose of the sanctions applied to heretics under canon law.[39] Only in the severest cases—as when a heretic was burned at the stake and their ashes thrown into a river—was the complete destruction of a person's memory the explicit goal; this practice also effectively prevented the collection of relics.[40] Yet in most cases heretics did not receive quite such drastic treatment; punitive measures for the most part focused on marking the disgraced person as separate from the community, in life and posthumously, rather than totally erasing any memory of that person. The sentence that was read to Galileo on 22 June 1633 is a case in point. It does not explicitly mention either his infamia or his memoria, most likely because Galileo was not found guilty of formal heresy but of the slightly lighter offense of vehement suspicion of heresy.[41] Yet, four measures imposed through the official punishment and as additional actions taken after the trial together tried to reshape the way Galileo was discussed and remembered during his lifetime and thereafter.

The first measure most explicitly relates to infamia. It entailed Galileo's short imprisonment at the Holy Office and his subsequent release to a mandatory period of house arrest at the Villa Medici, where he remained until the end of June 1633. He was then allowed to leave Rome and spend a few months in Siena, after which he returned to his own house in Arcetri, which had been his preferred place of residence for some years.[42] It is tempting to assume that the rapidity with which the formal imprisonment was transformed into house arrest demonstrates a leniency toward Galileo on the part of the Church, perhaps attributable to his connection to the Florentine grand duke. However, such an interpretation obscures the shame associated with formal imprisonment; as Masini's manual states, "Incarceration alone for the crime of heresy brings notable infamy to the imprisoned."[43]

While house arrest was not as shameful as formal imprisonment, both Galileo and his contemporaries saw it as a particularly severe punishment, precisely because—by literally keeping him at a distance from the city—it marked him as separate from the community to which he had previously belonged. This preoccupation with Galileo's visibility had in fact started at the trial's outset. Before he was transferred to the Holy Office, Galileo had stayed at the Florentine ambassador's residence in Rome, Palazzo Firenze, which was presented as a favor granted out of respect for his advanced age, his reputation, and his willingness to cooperate with the proceedings.[44] During his stay there he was not permitted to receive visitors or go outside, and even walks in the garden for exercise were forbidden. Only after repeated requests by the Florentine ambassador was he allowed to ride through the park of the Villa Medici in a half-closed carriage. The ambassador's many attempts to win Galileo permission to leave the embassy attest not just to Galileo's preference for exercise but also to the importance assigned by the various parties to the freedom of movement and the access to others that Galileo was, or was not, allowed.[45]

Interest in Galileo's visibility continued in the trial's aftermath. When Galileo petitioned to return to Florence, his request was only partially granted: He was allowed to leave Rome but not to return home. As a compromise, he spent some five months in the palace of the archbishop of Siena, Ascanio Piccolomini.[46] This period seems to have functioned as a quarantine of sorts—a test designed to monitor Galileo before permitting him to return to his former environment. The Inquisition had reason not to be entirely happy with the behavior of Galileo and his host, however. In February the Holy Office received an anonymous denunciation, stating that Galileo had

> sown in this city opinions not very Catholic, fueled by the Archbishop his host, who has suggested to many that he has been unjustly convicted by this Holy Congregation, and that he could not and did not have to condemn his philosophical opinions, supported by him with invincible and true mathematical arguments, that he is the first man in the world, that he will live eternally in his writings, even though they are prohibited, and that he is followed by all the best and modern [men]. And because these seeds from the mouth of a prelate may produce dangerous fruits, I hereby give an account of them.[47]

This report not only underlines the significance assigned to memory and posthumous glory in early modern Europe, but its reference to the archbishop's celebration of Galileo also gives us our first cue that the redirection of memory the Church was after did not materialize the way it had hoped. The report may even have spurred the Inquisition to adopt a stricter attitude toward Galileo. In November, several months before this report from Siena reached the Inquisition, Galileo had been granted permission to return to Arcetri (he was still strictly prohibited from entering Florence).[48] Within months Galileo again petitioned to return to Florence, but now the Inquisition resolutely refused and even warned him that if he asked again, they would have him imprisoned once more at the Holy Office.[49]

Galileo's many attempts to return to Florence indicate the value he assigned to the city and suggest that he experienced his confinement in Arcetri as a form of exile from Florence; his isolation there purposefully created literal and symbolic distance between him and his former social environment. He remained confined to Arcetri for the rest of his life, excepting a period of several months in 1638 after his request to return to Florence for medical treatment was granted. During these months, Galileo was even more severely cut off from society than he had been in Arcetri, perhaps in an effort to signal—to Galileo himself and more broadly to the Florentine community—that his return to the city in no way implied a return to grace. The restrictions on leaving the house were so stringent that they even severely affected his ability to worship, as he was allowed to attend Mass in the nearby church only if there was no crowd assembled there.[50] He was also not allowed to visit others, as a letter to the Florentine poet Buonarroti indicates: Galileo invited him to his house, as he had not been granted permission to go to Buonarroti's.[51] In Arcetri, too, Galileo could occasionally receive visitors, although not everyone was welcome to call on him and certain topics of discussion were strictly forbidden.[52] In a way, his informal exile at Arcetri was more effective than imprisonment in the Holy Office would have been. Galileo was far enough removed from the city to be invisible but close enough to cast a shadow, and as such his confinement and constricted visibility functioned as an immaterial monument to the Church's power.

A second measure the Church took against Galileo, one that concerned the prohibition of his *Dialogo*, similarly reflects this dual purpose of restricting access while not erasing the memory of Galileo altogether. The banning of books was a more moderate alternative to the practice of burning them, which, like the burning of heretics themselves, was a particularly heavy punishment mostly reserved for cases of formal heresy.[53] The decree that recorded Pope Urban VIII's decision with regard to Galileo's verdict suggests that the Inquisition in fact considered having Galileo's book burned. That decree includes, right after the title of Galileo's work, the words "publice cremandum fore" (to be burned publicly).[54] The words are crossed out and were consequently not included in the copies sent to the nuncios and inquisitors on 2 July, but there was nonetheless some confusion in Europe about the book's precise status. René Descartes was told that "all copies had been simultaneously burned in Rome" when he tried to buy the

Dialogo in Leiden and Amsterdam in November.[55] That the words were crossed out indicates the book was not to be condemned to eternal oblivion but instead should be remembered differently. The judgment marked the book as being both morally and factually wrong and—at least in theory—made it inaccessible; the prohibition forbade anyone to read the book without permission from the Congregation of the Index.[56]

The Church later sought to extend this measure and further restrict access to Galileo's work. As a result, all Catholic publishers were prohibited from publishing or editing Galileo's previous and future works. This, in combination with rumors that all his works were being removed from bookstores in Florence and Rome, led Galileo to conclude that "everything is done to remove memory of me from the world."[57] It is unclear whether there was a formal decision on future books or more of an informal guideline, as no formal decree stipulating a ban on the printing of Galileo's work has come down to us and as Galileo seems to have learned about it only through a friend in Venice who supported his attempts to publish his *Discorsi e dimostrazioni matematiche intorno a due nuove scienze*.[58] And, when Galileo eventually found the Leiden-based Elzeviers willing to print his new work in 1638, the book seems to have been available for sale in Rome, where approximately fifty copies were reportedly sold.[59] Galileo may thus have slightly overreacted. Yet, even the informal prohibition of the editing, publication, and perhaps even the sale of Galileo's books in 1635 does show that the Church, for a brief period at least, sought to restrict access to Galileo's work even further.

These two measures that aimed at immediately curbing the public memory of Galileo were complemented years later, in 1642, by a third one that specifically targeted long-term remembrance. That year, Pope Urban VIII and his nephew prevented the erection of a funerary monument to Galileo. Typically, in the immediate aftermath of a well-known person's death, there would be a public funerary procession, and a person of high status would deliver a laudatory oration. In a slightly longer time frame, a funerary monument or statue in a prominent public place might be considered. If the deceased's heirs were willing, the departed individual's final residence might even be transformed into a monument, an attraction for visitors who wanted to gain intimate insights into that person's way of life. Such monuments served not only to keep alive the memory of the deceased person but also to inspire others to greatness, as well as to strengthen local identity and reinforce civic pride.[60]

Galileo was given none of these signs of honor—but not for a lack of initiative on the Florentines' part. Immediately after Galileo's death on 9 January 1642, there was talk in Florence of constructing a tomb for Galileo in the Santa Croce church.[61] It appears there was also some doubt as to whether this would be possible, as a report was drawn up to document investigations into whether the construction of such a monument was allowed. The report argued that Galileo had not been declared infamous, as he had been found guilty only of vehement suspicion of heresy (rather than formal heresy) and that furthermore, by abjuring, he had been absolved of all punishments for this particular crime. The conclusion was that there was no reason not to honor Galileo with a funerary tomb.[62]

Yet a tomb for Galileo materialized only in 1737, as the pope and his nephew, Cardinal Francesco Barberini, successfully blocked the initial attempt.[63] A funerary tomb for Galileo, they thought, would not be appropriate and might also harm the Church's reputation. As the Tuscan ambassador was told, "It would not set a good example to the world if S.A. [the grand duke] were to do this thing, as he [Galileo] has been here in the Holy Office, for an opinion so false and entirely wrong, one that has also inspired many others there [in Florence], and caused a scandal so universal to Christianity with a doctrine that has been condemned."[64] The Holy Office also decreed that the grand duke should be advised that a monument would "scandalize the good."[65] Finally, the Florentine inquisitor was given further instructions for damage control that indicate the inverse relationship between Galileo's and the Church's reputations. In case the inquisitor proved unable to prevent the construction of a monument, he was to at least ensure that the epitaph did not harm the Church's reputation. The same terms applied to the funerary oration, which the inquisitor was to review before it was delivered or printed.[66]

The pope and his nephew chiefly relied on diplomacy and the goodwill of the Florentine grand duke to prevent the construction of a funerary monument. As the report discussed above also made clear, excluding formal heretics from receiving proper burial rites was standard practice, but this was not extended to those convicted of vehement suspicion of heresy. While the pope and his nephew were successful, their reliance on the inquisitor and on possible damage control strategies belies their lack of formal means to exercise influence over the shaping of Galileo's memory at this time. In the next section we will see that this is only one example of difficulties the Church encountered while trying to reshape the way Galileo would be remembered by future generations. Many of his contemporaries would try to find ways of celebrating him that did not openly oppose the Church's sentence but did not wholly concur with it either.

The fourth and final measure taken by the Church clearly sought to create and disseminate a new memorial narrative regarding Galileo, one that emphasized his penitence as well as the mistakes he had made as a scholar. Crucial in this regard was the circulation of Galileo's sentence and abjuration, ordered by Pope Urban VIII at the same time he decided on Galileo's final sentence.[67] Heretics normally abjured before a crowd; it was a custom that aimed to satisfy the betrayed Christian community's need for public atonement, emphasized the power of the Inquisition, and, more simply, set an example.[68] In a case of formal heresy, they would also be required to wear a *habitello*.[69] In line with the lesser severity of his crime, Galileo did not need to wear this humiliating piece of clothing, but he was required to publicly abjure and did so in the presence of eighteen witnesses.[70] On top of this, Urban VIII ordered that an additional and unusual measure be taken: Copies of the sentence and abjuration were to be distributed to all papal nuncios and inquisitors in Europe.[71] They were to pass them on to the specific group that the Church wanted to reach—namely, all professors of philosophy and mathematics.[72]

Copies of the proceedings sustained the sentence's visibility among a larger but also more focused audience than was usually the case for such punishments, in line with Galileo's fame in Europe. A letter by Cardinal Antonio Barberini to papal nuncios and inquisitors in all hubs of the Catholic world accompanied the sentence and briefly summarized the case against Galileo. It specifically stressed the warning he had been given in 1616, thereby emphasizing that Galileo had knowingly challenged the Inquisition's earlier verdict. The cardinal thereby aimed to set an example: "Knowing how the said Galileo has been treated, they can understand the seriousness of the error he committed and avoid it together with the punishment they would receive if they were to fall into it."[73] Finally, the inclusion of Galileo's abjuration challenged his credibility before his peers. It transformed the audience of mathematicians and philosophers into the virtual witnesses of an important event: the moment Galileo personally admitted that he had published and defended an opinion he now acknowledged to be wrong and promised to never hold again. Circulating both the sentence and the abjuration in this way strengthened the Church's evolving narrative that Galileo was someone who at first foolishly opposed its authority but then obediently retracted his erroneous position when told to do so. Such behavior marked him simultaneously as an unreliable scholar and a good penitent.

Through these four measures, then, the Holy Office sought to redirect the way Galileo was seen and remembered by his community and by future generations. The Church officials aimed to obscure the memory of his scholarly achievements and instead make him into a monument of their own authority. This did not mean they were immediately successful. The success of both infamy and memory sentences depended to a large extent on their intended audience's willingness to observe the punishing authority's directions for a reshaped memory.[74] As Charles Hedrick has pointed out, the sentences raise "fascinating questions about the relationship between elite political control and collective cultural tradition."[75] These questions have initially been taken up by scholars of antiquity but recently also gained prominence in the work of historians of early modern Europe.[76] A closer look at the way Galileo's supporters—in particular in Italy, where the reclamation was most directly felt, but also in other places across Europe—responded to the sentence reveals that they felt deeply uneasy with the Church's narrative.[77] What is more, they also found myriad ways to defy its authority and preserve their view of Galileo for future generations.

Galileo's Memory Makers

In the aftermath of the trial, Galileo's supporters still wished to celebrate him, but they felt hindered from openly doing so without restrictions. Reports drawn up in 1641 and 1642 show that there was some confusion over his legal status, even though the Holy Office's sentence had not explicitly mentioned infamia.[78] This state of uncertainty did not end with Galileo's passing in January 1642, and Torricelli's lecture, delivered in the summer of 1643, suggests that this confusion drove his Florentine contemporaries into a state of apathy. Without clear indications of what sort of celebrations would be permitted, they had given up after the initial attempts to construct a funerary monument were thwarted. Slowly but steadily even his supporters seemed to be forgetting Galileo. Torricelli, however, would not let that happen, and he was glad to be given the chance to remind his audience not just of Galileo's many achievements but also of their own role in safeguarding Galileo's memory.

The deaths of famous scholars had for centuries been surrounded with particular rituals. These efforts not only aimed to preserve, for future generations, memories of those whose achievements mattered the most but also gave students and others a chance to anchor themselves, albeit indirectly, to a tradition of greatness.[79] Torricelli cleverly drew on these ideas about the collective nature of glory to spur his fellow academicians into action. Shame, he warned them, would fall upon the academy if it were to go down in history with more infamous than famous members. Careful to avoid explicit mention of Galileo, the grand-ducal mathematician then turned to ancient examples. When someone famous dies, Torricelli said, those left on earth will form an image of him to conjure whenever they hear his name. They will pass this image on to future generations, but, as time passes, the image runs the risk of getting distorted. And thus, when hearing Nero's name, future generations might think of Augustus; instead of seeing an "impious, depraved traitor," they will see "a good, virtuous, faithful" figure.[80] Torricelli had found a clever, roundabout way of imploring his audience to defy authority and resist the way the Holy Office wished to portray Galileo, for if Galileo's own contemporaries did not recognize him for what he was, how could "those born a thousand years from now" remember him properly?[81]

Torricelli's lecture is one of many examples showing Galileo's supporters did not simply accept the Church's new memorial narrative about Galileo. Instead, Italian and European scholars showed their dedication to Galileo's memory by finding creative ways to laud him and his achievements in poetry, prose, and stone monuments—even if their celebrations were shaped by the need to balance their desire to openly, publicly, and explicitly praise Galileo and the need to fly under the Church's radar. Rather than give an exhaustive overview of every attempt to celebrate Galileo, the rest of this chapter explores five different dissimulation strategies Galileo's supporters developed in the wake of the trial to cultivate those elements of his fame they thought worth preserving for future generations.[82] My focus is in particular on early attempts to safeguard Galileo's legacy in the immediate aftermath of the trial, as these formed the basis for the better-understood late seventeenth-century and early eighteenth-century redirections of Galileo's memory, which constitute the final topic of this chapter.[83] Each of the strategies was shaped by specific cultural and political circumstances and is distinct in its foregrounding of elements of Galileo's memory and its defiance of the Church. All, however, aimed to change the memory culture around Galileo. At stake was the lasting recognition of Galileo's achievements, and his supporters, driven by a strong sense of justice, did what they could to ensure that their appreciation for his life and legacy survived into the future.

Questioning the Sentence

In the immediate aftermath of the verdict against Galileo, several of his supporters appealed to the Church to revoke its judgment. The best-documented attempt is the effort mounted by the French polymath Nicolas-Claude Fabri de Peiresc, whom we met in the second chapter of this book in regard to his desire to rename Jupiter's satellites. In 1634 and 1635, he wrote to Cardinal Francesco Barberini in Rome and urged him to discuss with his uncle a possible lightening of the sentence.

Peiresc's letters adopt the perspective of future generations and explicitly recognize the importance assigned to Galileo's memory for the long term. Early in the first letter, Peiresc expressed hope that the cardinal would do "something for the consolation of a good old septuagenarian, who is in ill health and whose memory will be difficult to erase in the future."[84] He expanded on this point some lines later, almost casually reminding the cardinal of

Galileo's many achievements: "It will be difficult for posterity not to show him eternal gratitude for the wonderful novelties he discovered in the heavens by means of his telescope and his extremely penetrating mind."[85] Peiresc thus suggested that the attempts to obscure Galileo's achievements would not work and implied that the Holy Office should revoke (parts) of the sentence to align itself with the verdict of future generations.

Peiresc then pointed out the extent to which memories of Galileo and the Church had become intertwined. This was precisely what the Holy Office had been after: Galileo's defeat was to be firmly linked to the Church's victory. However, Peiresc claimed that the Church's efforts would produce the opposite effect because the Church had treated Galileo too harshly. To highlight this point, Peiresc first provided several examples from two categories of sinners who (unlike Galileo) had been shown mercy: Church fathers and artists. These groups include men who "fell into some errors" (the Church fathers) or committed "extremely serious sins whose enormity was most horrifying" (the artists). They were nonetheless pardoned by the Church, either because they showed other signs of piety (the Church fathers) or "in order not to let their merits go to waste" (the artists).[86] If so many sinners have been shown mercy, why not extend the same leniency to Galileo, who was old and frail? Future generations, Peiresc emphasized, "may find it strange that . . . so much rigor should be used against a pitiable old septuagenarian."[87] He then finally came out and said that the pope's and the Church's memory, rather than Galileo's, would suffer in the future: "This will really be deemed most cruel everywhere, and more by posterity, for it seems that in the present century everyone neglects the interests of the public (especially of the disadvantaged) in order to look after his own. And indeed it will be a stain on the splendor and fame of this Pontificate, if Your Eminence does not decide to take some precaution and some particular care."[88]

The letter did not have the desired effect, but Peiresc did not give up easily. In his second letter, he reminded Cardinal Barberini of the fate of another scholar who had been convicted of offenses in his lifetime but was then honored by posterity: Socrates.[89] Peiresc first pointed out that a pardon for Galileo would surely elicit the approval of "the noblest minds of the century," to then contrast this positive effect with the alternative: "However, a contrary outcome would run the risk of being interpreted and maybe compared someday to the persecution of the person and wisdom of Socrates in his own *patria*, so much reproached by other nations and by the very descendants of those who caused him so much trouble."[90] Peiresc questioned not only the sentence's justness but also its effectiveness by predicting the responses of "the noblest minds of the century" as well as those of future generations.

His is the most clearly formulated critique of the sentence we will encounter, and Peiresc paired this explicitness with a direct yet private way of communicating it. Peiresc voiced his criticism in private letters addressed to one of the cardinals who had led the investigation into Galileo's beliefs and works. Such directness was made possible by the privacy of the medium: Unlike Galileo's letters to Castelli and the grand duchess, these missives were not meant to be shared with others. Other factors also enabled Peiresc to be so frank in his criticism: His socioeconomic status, his large network of influential peers, and his country of residence (France, where the Inquisition's verdict had no jurisdiction) put him in a relatively safe position from which to criticize the Church's verdict in such a direct manner.[91] Peiresc also had a prior connection with Barberini, whom he—like many modern historians—may have suspected of nursing private sympathies for Galileo: Barberini had been a member of the Accademia dei Lincei since 1623 and was one of the three cardinals who did not sign Galileo's sentence in 1633.[92] Still, Peiresc's pleas did not have the desired effect. Cardinal Barberini promised to discuss the subject with his uncle, but no concrete measures were taken, and Peiresc did not raise the issue again—either in private or in public.[93]

Covert Criticism

Peiresc chose to voice his criticism directly but privately. Two other supporters did precisely the opposite when constructing monuments—one literary, the other material—for Galileo. They publicly criticized the Church's verdict and did so in a much more indirect and oblique way.

We first turn to an epigram written by the French scholar Gabriel Naudé. Naudé, who studied with Galileo's old colleague Cesare Cremonini in Padua around 1620, excelled in the art of dissimulation—the concealment of one's true feelings and intentions to avoid censure.[94] Upon moving to Rome, he also struck up a fruitful collaboration with the famous Roman antiquarian Cassiano dal Pozzo, who served as secretary to Francesco Barberini and was, like Galileo, a member of the Accademia dei Lincei.[95] Among Dal Pozzo's many possessions was a substantial collection of portraits of illustrious men, and early in 1641 Naudé published a book of epigrams celebrating some of them.[96] It included a seemingly innocuous poem in honor of the Tuscan scholar:

> It is not your face, Galileo, I care to see,
> But the painted picture with your eyes pleases me all the more:
> For with your eyes I saw somehow the stars revealed,
> and the justice administered by you to a new heaven.
> Now I could not behold with a calm mind your eyes,
> which were hidden a while ago under the blind night.[97]

At first glance, Naudé's epigram simply seems to offer modest praise of Galileo, whose eyes are first celebrated and then mourned, as an illness had rendered him blind several years prior. But upon closer inspection the poem, which explicitly draws on the tension between visibility and invisibility, tells a different story, revealing a clever criticism of the trial hiding in plain sight.

Naudé achieves this by making full use of the ambiguous meaning of various words. In the fourth line, he praised Galileo because he has administered justice to a "new heaven." The word *iura* infuses his poem with a juridical connotation that brings the Inquisition's verdict into memory. But Naudé has reversed the outcome, presenting Galileo as the judge administering justice to the cosmos rather than as the subject of a juridical procedure. This implies that Naudé thought Galileo's discussion of heliocentrism had been just, and yet, if anyone protested, plausible deniability would allow him to claim he had been referring to some other discovery by Galileo, such as the satellites revolving around Jupiter.

The final two lines express Naudé's uneasiness when looking upon Galileo's portrait, which he "could not behold with a calm mind." In the most straightforward reading of these two lines, Naudé simply mourned Galileo's tragic loss of eyesight. But "I could not" might also refer to an injunction or prohibition; with the Inquisition sentence in mind, Naudé could feel as if he, regrettably, was no longer permitted to admire Galileo. The "nocte caeca" under which Galileo's eyes have been hidden similarly lends itself to different interpretations. Literally "a blind night," the phrase "nocte caeca" within a poetic context can also be translated as "obscurity" or "ignorance."[98] Have Galileo's sharp eyes, which are so strongly linked to his achievements in the poem's third line, been prevented from seeing more by the ignorance of those who sentenced him and wished to cloud his achievements in obscurity?

The poem's ambiguity allows it to be read either as a modest and appropriate appraisal of Galileo or as criticizing the trial for causing damage to the memory of Galileo and to the progress of scholarship. Galileo certainly seems to have gotten the message: Immediately upon receiving the book of epigrams, he wrote a jubilant letter to Cassiano dal Pozzo expressing his gratitude for "the gain of my reputation for the world."[99] It is more difficult to establish whether others, especially the officials of the Church, had been as perceptive and picked up on Naudé's wordplay as well. Cardinal Francesco Barberini was certainly well positioned to do so, given his proximity to the work, Naudé, and Dal Pozzo. The portraits Dal Pozzo had collected also included one of the cardinal, and Naudé's book of epigrams accordingly included a poem celebrating him, as well as one praising his uncle, the pope. The epigram to Francesco Barberini specifically references the cardinal's library; Naudé would later, in 1641, be appointed its librarian.[100] Had Barberini picked up Naudé's book and read the epigram dedicated to him, he would have had to turn a mere two pages to see the epigram in honor of Galileo.[101]

The material monument Vincenzo Viviani, Galileo's last disciple, erected for Galileo in 1693 similarly criticized the Church's narrative in a covert and indirect manner. In the decades after Galileo's death, Viviani had become ever more pessimistic about the prospect of erecting a funerary tomb for Galileo in the Santa Croce church.[102] He resolved to instead embellish his own residence with three large plaques or scrolls dedicated to Galileo's memory and, in doing so, to create a monument that, constructed on Viviani's personal property yet visible to all passersby, is at once private and public. Aware of the value of such a monument to the wider community of European scholars, Viviani published the text

Fig. 5.1. Vincenzo Viviani's Palazzo dei Cartelloni celebrates Galileo's life and achievements for everyone in Florence to see. Source: Wikimedia Commons.

for the benefit of international readers, and his house, known as the Palazzo dei Cartelloni (see fig. 5.1), became a "site of intellectual pilgrimage."[103] On the three scrolls, pilgrims would find an enthusiastic celebration of Galileo's many achievements and his links to Florence. But Viviani also daringly included several lessons that he had learned from Galileo, including "the need to defend firmly truth and justice" and the unfortunate lesson "that lies, flattery, and hypocrisy should be avoided like the plague."[104]

These statements take up only a minuscule part of the three scrolls devoted to Galileo's memory. As in Naudé's epigram, there is no explicit reference to the trial, and the implicit message in the statements must have been sufficiently concealed to go unchallenged by the authorities. Viviani nonetheless takes Naudé's criticism one step further, praising and even encouraging precisely those characteristics of Galileo that had landed him in hot water and raising the question of *whose* lies, flattery, and hypocrisy should have been avoided. The answer was most likely those of Urban VIII. As we have seen in chapter 3, that pope had once sent Galileo a poem called "Destructive Flattery," which was printed and reprinted several times in the course of the seventeenth century.[105] This version of events challenges the notion of Galileo being a good penitent and a testament to the Church's authority. Here, rather than meekly accepting the Church's reasonable verdict after seeing the errors of his ways, Galileo is cast as firmly defending the truth and inspiring others to do the same.

Naudé's epigram and the façade of Viviani's palazzo aptly demonstrate that within the Catholic world, criticism directed at the Church's treatment of Galileo could publicly be voiced so long as it was covert enough and allowed room for an alternative reading. The creators of both monuments, purely by avoiding any explicit reference to the trial, smuggled veiled criticism of its outcome and value into the public sphere, winning a subtle yet important victory.

Explicit Compliance, Implicit Resistance

The third strategy uses a show of explicit compliance with the Inquisition's verdict to implicitly resist it. This strategy is evident in three sonnets dedicated to Galileo by Paganino Gaudenzi, a Swiss-born, Protestant-turned-Catholic professor of philosophy and literature at the Pisan studio. Gaudenzi composed the sonnets immediately after Galileo's death in January 1642 and likely printed them on his own press.[106] The poems themselves are not particularly creative or interesting and in many ways are quite similar to the ones discussed in chapter 3 of this book: They praise Galileo's achievements, laud his lynx-like vision, and contrast the cruelty and brevity of life with the eternal rest of death. However, Gaudenzi felt that they needed to be followed by a disclaimer, and he stated that his praise for Galileo did not mean he approved of "the mobility of the earth, Academically proposed by him in his Dialogues, or another opinion that could be contrary to the determination of the Holy Church."[107]

Gaudenzi's work thus pairs praise for Galileo's accomplishments with an apparent genuflection to the justness of the Church's sentence. There are clues, however, that this conspicuous nod of approval toward the sentence was not entirely sincere and might even be read as ironic or satirical. When Gaudenzi sent his poems to his correspondent Cassiano dal Pozzo in Rome, he requested that he not share them with anyone, "for the reasons you know."[108] Possibly, Gaudenzi feared that his sonnets, which he claimed were to the taste of the grand duke, would not be equally appreciated by Dal Pozzo's patron, Cardinal Barberini. Gaudenzi was not being overly careful, since attempts to honor Galileo were occasionally thwarted by the Inquisition. A good example is that of Jean-Jacques Bouchard, who in December 1637 delivered a eulogy dedicated to the memory of Peiresc before the Roman Accademia degli Umoristi. As was common, Bouchard emphasized Peiresc's many learned friends and praised Galileo among them as prince of the mathematicians "without dispute" (*Mathematicorum sine controversia Princeps*).[109] Bouchard had this eulogy printed in Venice and sent a copy of it to Galileo via a mutual friend.[110] So far, all was well—but then Francesco Barberini initiated the publication of a literary commemoration in honor of Peiresc in Rome. The Barberini family asked Bouchard to contribute his eulogy to this work and then told him he must replace his reference to Galileo with something less controversial.[111] Bouchard obliged but later wrote that these "barbarianisms used against poor Galileo" inspired him to dedicate his next bit of spare time to writing a biography of Galileo (he never did).[112] This episode highlights the obstacles encountered by those devoted to Galileo's memory as well as the limits of the Church's power. A remark as seemingly innocuous as Bouchard's needed to be deleted for a work printed in Rome, even as the original version of Bouchard's eulogy had already been published in Venice and was reprinted in Aix-en-Provence a year later.[113]

Gaudenzi continued to try and find a balance between doing what he felt was right and doing what was allowed. Lines from an earlier (1635) sonnet of his, which mourned the death of the Pisan mathematician Niccolò Aggiunti, further suggest that his disclaimer for the 1642 poems should not be taken at face value. Now that Aggiunti had reached heaven, Gaudenzi reasoned, he could avail himself of a better view of the celestial bodies. This meant he could "now see, if the immense sphere of the sun / Is the base of the world and the center, and if the moon / Keeps valleys, waves, and mountains inside its globe."[114] The lines suggest that for Gaudenzi the debate about the nature of the cosmos was by no means closed, regardless of what the Catholic Church had decided. This impression is further strengthened by a statement made in his *La Galleria dell'inclito Marino* (1648).[115] This work included the sonnets in memory of Galileo, but this time they were preceded by a brief introduction that served as a useful reminder of Galileo's many achievements as well as his favored position at the Florentine court. Gaudenzi hinted at the ongoing discussion of the heavenly bodies and linked it to Galileo's eternal glory: "Hence if the nobility and greatness of the subject provides reputation and esteem to the practitioners of the sciences, he above all others in this time of celestial contemplations may be called worthy, and one may truthfully say, that as long as the sun, and the other planets, are talked about, bright and illustrious the memory of Galileo will be."[116]

Six short years made a crucial difference in the freedom Gaudenzi allowed himself. In 1642, he had restricted himself to printing the sonnets and the accompanying disclaimer on his private press, circulating them only among close acquaintances; in 1648, he had the sonnets published and allowed them to circulate freely, even adding further commentary on Galileo's lasting memory. This change in attitude may have been prompted by Pope Urban VIII's death in 1644, which also led his nephews Antonio and Francesco to flee from Rome. Urban's successor to the papacy was Innocent X, his former rival Giovanni Battista Pamphili. The new pope would certainly not tolerate any slights on the papacy itself, but authors may nonetheless have felt they had slightly more room to maneuver under his rule.[117] Still, even in 1648 Gaudenzi took care not to push his luck. His phrasing remained cautious, and the sonnets were again accompanied by a (somewhat toned-down) disclaimer.[118]

In 1642 as well as in 1648, Gaudenzi chose to comply explicitly with one element of the Church's preferred memory for Galileo. Doing so created space to praise Galileo's achievements and to include a cryptic comment on Galileo's long-lasting glory, which derived from precisely those astronomical discoveries deemed problematic by the Church. As such, Gaudenzi's work offers an example of resistance disguised as compliance with the officially sanctioned memory culture. As long as one openly paid lip service to the verdict that had rendered Galileo's discoveries problematic, it was possible to celebrate Galileo's achievements and to hint that someday in the future his accomplishments would be regarded quite differently.

The Narrative Subverted

Vincenzo Viviani wrote his life of Galileo, *Racconto istorico della vita di Galileo*, in 1654 at the request of Prince Leopold de' Medici (brother to Grand Duke Ferdinando II).[119] The text aptly exemplifies the trial's impact, as even Viviani, one of Galileo's most fervent admirers, could not openly laud every aspect of his life and career in a genre designed precisely for this sort of hagiographic praise.

Viviani's account echoed Florentine models of greatness and closely mirrored Giorgio Vasari's 1550 life of Michelangelo.[120] However, the two biographies starkly differ in their discussion of the role of divine intervention. Vasari invoked such intervention to underline Michelangelo's greatness, opening his account by proudly stating that God had sent Michelangelo, a true universal genius whose eminence would surpass that of all previous artists.[121] Viviani, for his part, brought in the divine in an attempt to remove the stain on Galileo's memory. The mistakes Galileo made in publishing the *Dialogo*, Viviani wrote, were God's way of allowing him to show his humanity: "But while Galileo had already reached as far as the heavens with immortal fame won by way of his other admirable speculations, and acquired a divine reputation among men with many new things, Eternal Providence allowed him to show his humanity by erring."[122]

Viviani thus presented Galileo's mistake and the subsequent trial as favors from God: They were obstacles that Galileo needed to overcome in order to become truly heroic.[123] This way, Viviani stretched the narrative of Galileo as a good penitent as far as possible, inverting the Church's verdict so as to make Galileo a good scholar once more. Precisely because Galileo had erred, he then learned to recognize his own mistakes, which transformed him into the remarkable, memorable scholar he became. If this narrative already seems unconvincing, it is further undermined by a small detail in Viviani's account. Directly after presenting

Galileo's erring as a divine favor, Viviani depicted the Inquisition's decision to send Galileo to Siena as a good thing, because it helped him avoid the plague that was afflicting Florence. Viviani similarly framed the Holy Office's decision to confine Galileo to Arcetri as positive, since the town had "good air." It was also "conveniently near the city of Florence," which allowed for visits from friends and family offering "relief and consolation." Though the only allusion to the adverse situation that Viviani permitted himself, the acknowledgment here that Galileo needed such support, is enough to let us question Viviani's previous interpretation of the trial as a blessing.[124]

The story is wobbly and also seems to have left Viviani unsatisfied; he never published this account in his lifetime. Earlier studies have interpreted the delay in publication as due to its author's perfectionism, but it is also possible that Viviani might have felt that this particular account did not do Galileo's memory justice.[125] As we have already seen, Viviani later recalled that Galileo had taught him "the need to firmly defend truth and justice." This lesson stands in stark contrast to his tone here, especially to his portrayal of Galileo piously welcoming his punishment. Time, it seems, may have played a role. Near the end of the century, Galileo's conviction was not as sensitive a subject as it was in 1654, twelve years after his passing. Another possible reason for the difference in tone is that Viviani wrote his *vita* of Galileo at the behest of Prince Leopold, to whom it is addressed as a letter—and as such, Viviani needed to comply with the more placatory Medici memorial strategy.

Partial Forgetfulness

The last strategy for dealing with the public memory of Galileo after the trial is especially visible in memorials with a more official character, such as those erected under the patronage of the Medici family and the funerary tomb eventually constructed in the Basilica of Santa Croce in Florence. These combine praise for Galileo's achievements with complete silence regarding the trial.[126]

The grand duke had maintained contact with Galileo and visited him several times in Arcetri after the trial in 1633. Galileo interpreted this as an important sign in his favor and an indication that his reputation was not completely lost after all.[127] Moreover, Leopold de' Medici asked Galileo in 1640 to respond to the work of Fortunio Liceti, thereby signaling that he wished Galileo to continue his work as the grand duke's First Mathematician and to publicly serve the family.[128] As for material monuments, however, the Medici were less forthcoming. Perhaps feeling bound by the Church's request for discretion, they initially preferred to restrict their material celebrations of Galileo to the private sphere: When they honored Galileo by commissioning a portrait of him from Justus Sustermans around 1640, they kept the painting in the private apartments of their Palazzo Pitti residence.[129] The boundaries between private and public are somewhat blurred here, and the grand ducal family's display of the portrait in their private apartments not only signaled the Medici family's great esteem for Galileo but even implied an enviable level of intimacy. However, the lack of public monuments to honor Galileo did not go unnoticed by supporters like Viviani, and some twenty years would pass before the Medici took a first small step toward a more public sign of honor by hiring Agnolo Gori to depict Galileo, champion of mathematics, on the ceiling of one of the Uffizi's corridors.[130] In 1674, the Medici took an additional step by moving the portrait by Sustermans from their private apartments to the Uffizi.[131] The portraits, like all private and public monuments the Medici dedicated to Galileo, avoided all references to the trial and instead foreground his scholarly achievements.

The same is true for the tomb in the Santa Croce basilica, which eventually materialized in 1737. It was, given its place in Florence's unofficial pantheon of illustrious men, the most formal, public, and authoritative monument to Galileo's glory. The story of the monument, and Galileo's reburial, signals a crucial new development in the remembrance of Galileo. On 12 March 1737, Galileo's body was moved from the remote, solitary little room in which he had first been entombed to his new grave in the basilica's central nave. He was given pride of place opposite Michelangelo, whose bones had been smuggled from Rome to take their proper spot in the Florentine pantheon. The tomb is as much a monument to Galileo as to his supporters and captures the collective nature of posthumous glory as no other: A large part of the epitaph is devoted to acknowledging the individuals who helped, at last, to construct the long-awaited tomb. In particular, the construction of Galileo's final funerary tomb would not have been possible without the continued devotion and financial support of Viviani, who after his death left a considerable sum for its execution—on the condition that he would be buried with his former teacher.[132] This classic form of veneration became an essential element of the tomb itself. The portion that specifically concerns Galileo, on the other hand, is modest and succinct. It praises Galileo as a great "restorer" of geometry, astronomy, and philosophy and makes no mention at all of the trial.[133]

That the trial receives no mention on the official monument is remarkable. This aspect of Galileo's life was by no means forgotten, and in fact the Florentines chose to explicitly commemorate it in the Santa Croce church, on a separate plaque placed near Galileo's first, more modest grave.[134] The epitaph communicates the feelings of frustration and sorrow apparently widely shared in the city: It states that all cultured men and visitors of Florence have seen Galileo "lying here without honor but not without tears" and that "all men of good conscience are happy that he has finally been moved from here to a more worthy resting place."[135] Much more emotional in tone than the official monument, the plaque shows that the Florentines did not want to forget Galileo's struggle but actually chose to memorialize their grievance over the injustice done to Galileo and his memory.

Fear of the Church's authority may have played a role in the choice to capture this sentiment on a separate plaque rather than on the official tomb, even as the relationship of the Church to the Tuscan city had changed in the years before Galileo's body was moved—the election of the Florentine-born, Pisa-educated jurist Pope Clement XII in 1730 was especially encouraging.[136] Still, Urban VIII had also been welcomed enthusiastically as a Florentine pope, and expressing frustration about the trial via the alternative epitaph meant that the plaque bearing that text could, if the need arose because of changing politico-religious circumstances, be removed without damaging the official tomb. But the Florentines may also have hesitated to include the more negative elements of Galileo's memory on the official tomb, either out of shame or more likely out of a sense of defiance stemming from the idea that Galileo had been wronged. This tomb marked Galileo's admission into Florence's most visible and recognizable gallery of honor, and perhaps the Florentines simply thought this was not the place to recount any aspects of Galileo's life that would cast a shadow on his memory. Even a modest reference to Galileo's innocence and the unjustness of the verdict would remind people of the trial, thereby effectively directing the focus away from Galileo's stellar achievements.

The omission, then, can be regarded as an intentional stance of dignified silence, an attitude that refused to let Galileo's memory be sullied by a narrative that had been marked as shameful, at least by others, and that could not be set straight in the here and now. The new tomb constitutes an attempt to create a new public narrative for Galileo—one that focused solely on his achievements and celebrated him in the same way as that other marvelous Florentine, Michelangelo. As such, it entirely flips the preferred memorial narrative of the Catholic Church, which tried to obscure Galileo's achievements and foreground his failures.

Meanwhile, the reburial also gave a new impulse to a narrative that has come to overshadow the one that solely remembers Galileo for his achievements. The reburial itself was an austere but highly orchestrated event. It was recorded in detail by the Società Colombaria Fiorentina, the newly founded society that aimed to preserve Florence's glorious past and that invited its members and a notary to witness this key moment in Florentine history. A group of men consisting of the basilica's priests and several of Florence's most prominent noblemen, intellectuals, and artists moved Galileo, Viviani, and the female body they discovered to be buried alongside Galileo (which was later assumed to be one of his daughters, Suor Maria Celeste) to their final resting place opposite Michelangelo.[137] The men could not restrain themselves and famously took home three fingers, a tooth, and a vertebra, which they put on display for friends to admire. As relics, these bits of Galileo underlined his status as a scholarly saint and martyr who suffered for his beliefs.[138] This particular narrative, which emphasizes Galileo's achievements and his wrongful suffering at the Church's hands, came to be especially prominent in the eighteenth century as the Church's influence in Italy lessened and nationalistic impulses inspired stories of secular Italian greatness. It gained further relevance in the nineteenth century and is present to this day in popular culture.[139]

As we have seen, however, elements of this narrative had already cautiously been developed in the seventeenth century; we can detect its presence in the work of Naudé, who portrayed Galileo as the victim of a legal error, and in Viviani's depiction of Galileo as being tested by God. Both men, alongside a host of other Galileo supporters, had to tread cautiously in the immediate aftermath of the 1633 trial. This was its express purpose and in line with the way heretics were treated under canon law. Although the Inquisition did not condemn Galileo to full, legal infamia after he abjured and cursed his errors, it did target his broader social fama as well as his memoria—the way he would be remembered in the future. Through formal and informal measures, the Inquisition aimed to restrict Galileo's posthumous glory by marking his scholarly achievements as wrong and instead shaping him into a monument to the Church's authority.

Galileo's supporters clearly felt the weight of the Church's initiatives but persisted and found ways to celebrate Galileo while avoiding censure. His memory makers did not know precisely what would or would not be allowed, and in the years immediately following the trial, as in the decades after Galileo's death, they cautiously set out to find out how far they could go. Occasionally they were rebuked, but direct confrontations were rare, and most supporters appear to have nimbly flown under the Church's radar. They did so partly by eschewing explicit references to the trial and the way it had sought to damage Galileo's glory, instead choosing to emphasize Galileo's achievements. Those who went a step further and combined praise for Galileo with mild criticism of the trial usually phrased their objections in such a way that they could read it in a different manner. One supporter even created room for more abundant praise of Galileo by paying lip service to his conviction. Each found a different balance between directly or indirectly, publicly or privately, and covertly or overtly resisting various strands of the memorial narrative the Church sought to establish, but all contributions share appreciation for Galileo's achievements, empathy for his tragic fate, as well as a belief that Galileo's story did not have to end with the trial or his death.

In this, they—and especially Peiresc—have been proven right. While the Catholic Church has been extremely successful in making sure that Galileo's memory is forever tied to the trial, future generations have mostly found the Church's verdict to have been too harsh. They have, moreover, not forgotten Galileo's many achievements, but that Galileo is remembered today for the trial as well as his many achievements is ultimately the result of the efforts of his many supporters—those who realized they had a role to play in the way Galileo would be remembered by generations to come.

Epilogue

Legacy

OVER THE PAGES OF THIS BOOK, WE HAVE MET GALILEO IN many different guises. Starting with Galileo as the elusive and somewhat dubious witness in a thorny inheritance dispute, we then moved to the young, scrappy, and ambitious mathematician Galileo, who was well connected but struggled to find jobs nonetheless and was rejected from a university teaching position for his lack of fame. Subsequently, we met Galileo the accused plagiarist, dependent on friends, patrons, and the university board for his defense, eager to broadcast their words of praise in print. Suspicions nonetheless lingered, and when all of Europe sought to determine whether Galileo really was the first to observe four satellites around Jupiter, he simultaneously became a symbol of both Florentine glory and Florentine deceit. We have also encountered Galileo as the heroic discoverer of new worlds, praised and celebrated for his astronomic findings and the use of the still new and controversial telescope; he is one and the same Galileo others saw as a firebrand innovator and leader of a group of rebellious, reckless Galileisti roaming the streets of Florence. Finally, we have followed the semiforgotten penitent Galileo from Rome to Siena and Arcetri and have seen how he was gradually transformed, following the continued commitment of Italian and European supporters, into Galileo the heroic but tragic martyr of science.

Some of these versions of Galileo are much more familiar than others, but the point of this book has not been to expose the "true" Galileo among the various contenders. Rather, it has sought to show that Galileo meant different things to different people—

from the very start of his career to long after his death. Galileo's fame was the direct result neither of his own achievements nor of his tactical maneuvering: To gain widespread recognition for his discoveries Galileo was almost entirely dependent on the efforts of other people—the arbiters, brokers, or mediators who collectively invested time, energy, and resources into continuously and expressly cultivating, shaping, and disseminating his fame. Those people are easily overshadowed by the person who obtains fame, but studying their actions is vital if we aim to understand the particular local and cultural contingencies that lay at the foundation of narratives of individual greatness. This lens also brings into focus the mechanisms through which fame is formed—the competition, appropriation, and opposition fame spurs among different groups, who in turn shape it according to their hopes, fears, and desires.

The continuous negotiations over how Galileo should be seen directly shaped his fame and career. They also point to wider debates about the value of new knowledge, new forms and practices of knowledge production and dissemination, and the changing role of scholars and their audiences. Galileo's fame formed a focal point for the tensions these changes evoked, and this book showcases the view that if we want to understand the recognition bestowed on individual scholars in a particular period and time, we need to study the complex interplay between cultural, religious, and legal norms and practices on which their contemporaries leaned. Though highly educated and members of the cultural elite, many of Galileo's fame makers are not traditionally associated with science and the people who practice it. But if we want to fully grasp the particular and quickly changing world of early modern science, we also need to study the preachers, poets, and painters who were impressed, intrigued, or offended by novel claims and new discoveries. In a period when the boundaries between different disciplines and art forms were more porous than they are now, when mathematicians wrote about the dimensions of Dante's hell, painters were inspired by new astronomic findings, and masked philosophers, Jesuit poets, and Dominican preachers playfully or viciously debated the value of novel findings and instruments, their voices vitally shaped the new role of scholarship in society and made fame at once a highly coveted and a highly ambiguous asset.

Drawing on cultural, humanist frameworks that linked extraordinary achievements to glory, Galileo's supporters not only presented his growing fame as the rightful reward for his merit but also further promoted his novel findings as heroic and adventurous contributions to new scholarship and new ways of investigating nature. Fame's affective appeal and collective nature spurred these supporters to engage with Galileo's fame, not only in manuscript but increasingly in print as well. To them, Galileo's fame was an inherently valuable asset that represented an opportunity to connect with people farther away and attract students or readers across Europe. They eagerly took it upon themselves to further enhance his fame, fully aware that this would benefit themselves as well. Others, however, had a decidedly more negative view, and discussions about Galileo's discoveries quickly became enmeshed with debates on the morality of his fame. The attacks by Galileo's Aristotelian and Dominican criticasters, steeped in the Christian language of virtue and vice, posited that the desire for fame made scholars lazy, sloppy, careless, or even dishonest, as scholars like Galileo pursued and defended ideas they knew to be false. Chasing novelty for novelty's sake, they willfully defied established authorities, whether scholarly or ecclesiastical. To these adversaries, Galileo's command of print was especially suspicious, as this significantly broadened the potential audience for his novel and controversial ideas and made it impossible to limit their circulation to experts who would be able to properly assess the value of his claims. Galileo's opponents held that most people lacked the wisdom and experience to see through the illusion of merit that fame created, and therefore they might all too easily be persuaded by his claims. In the eyes of these opponents, this was especially problematic, as large groups of people were not just to be protected against novel ideas but also could turn into fearsome and uncontrollable masses. Such fears rested on the widespread idea, held not only in politics but also in courtrooms, that large, anonymous groups of people and the talk shared among those people easily spurred disorder, a dangerous loss of control, and social, political, and religious upheaval. These suspicions of crowds, talk, and rumor and print intersected in Galileo's case, as he attracted the attention of readers abroad as well as students in the local settings in which he and his direct adversaries moved, and that helps explain why they objected not just to his claims but to his fame specifically. Both the excitement and the unease Galileo's fame elicited, then, can ultimately be traced to questions about which achievements were worth celebrating and who possessed the ability and power to decide what constituted good scholarship when scholarly findings and discoveries increasingly captured the attention of a much wider audience than before.

Regardless of their intent, Galileo's opponents further enhanced his visibility by responding to his work in print and from the pulpit. The ensuing debates about the precise value of his fame created momentum for a novel role of scholars as highly visible public figures whose work appealed to wider segments of society. In highlighting this development, this book has drawn on scholarship that traces the rise of modern science to the increased circulation of ideas in the early modern period, which, historians argued in the last decades, brought about the particular norms of disinterested, open exchange and collaboration that characterize modern science.[1] The story of Galileo's fame complicates this picture, as the increased interest in scholarly ideas and the increased public presence of scholars did not necessarily give rise to rational, free debate, even among those who saw fame as a positive asset. Instead, fame sparked and fueled collaboration as well as fierce competition, controversy, and conflict; it divided as much as it united, showing the fraught and fragmented nature of the public sphere as it developed.

Nowadays, there is little debate whether scientific discoveries should be shared with wider audiences; if anything, most scholars call for greater accessibility and transparency. But while fame is no longer as contested as it once was—if anything, it is associated with superficiality rather than with danger—the fame of individual scholars remains a divisive topic. Though Galileo's fame makers were keenly aware of fame's collective nature, its growing importance has actually led to an increased emphasis on individual merit and achievement that informs scientific culture to this very day. This emphasis on individual merit, however, does not go uncontested, and discussions about the best way to recognize scholarly achievements continue within and outside of academia. These debates are especially poignant, as the increased public presence of scientists has also, ironically, increasingly made them figures of political interest precisely because of the associations with the free and rational pursuit of truth that science evokes.

Unsurprisingly, Galileo remains a focal point for such debates. When the history of science emerged as an academic discipline in the twentieth century, its earliest practitioners focused especially on isolated individuals, tracing a clear genealogical line from the brightest minds of the past to what they held was a particularly modern way of thinking. Having been directly influenced by the politics of World War I, the subsequent rise of totalitarianism, and eventually the Cold War, they nonetheless contrasted the rationalizing impulses of science with the "corrosive influences of irrational ideology" and set science and scientists apart from the rest of society, painting them as "free, rationalizing individuals" unconcerned with politics.[2] They were not the first to do so, nor were they the first to turn to Galileo as a symbol of their claims: The story of Galileo's struggle with the Catholic Church gained new momentum with the emergence of Italian nationalism in the nineteenth century, when the unification of Italy spurred anti-Catholic sentiment. It did so again during the period of Italian fascism, when relations between Benito Mussolini and the Catholic Church soured. In 1942, Mussolini was rumored to have personally assisted in purchasing Galileo's villa in Arcetri for the Italian state.[3]

In recent decades, historians have carefully sought to dismantle the oversimplified narratives of progress that make scholars such attractive symbols for political movements. They have not only made it clear that Galileo's trial was not a simplified, black-and-white story of a genius maverick working against the religious machine but also paid careful attention to the work of artisan practitioners and invisible technicians. By highlighting the collaborative nature of scientific discoveries, they have shown that it makes little sense to focus on only a few illustrious, extraordinary figures. However, the impact these insights have made on the wider public seems to be more limited, and familiar stories of individual genius remain popular outside of works aimed at specialized historians with years of training behind them. It is not difficult to understand why: They offer a comfortable view of history in terms of linear progress as well as easily recognizable focal points around which to organize broader narratives of scientific discovery. Moreover, stories of scientific genius are firmly embedded in national history curricula that tend to foreground contributions from the nation's "own" historic figures.[4] In today's world, they are also mirrored in media reports that pair stories of individual research breakthroughs and "Eureka" moments, with interviews spotlighting scholars' personal lives, habits, and interests. The commercialization of science and celebrity here go hand in hand and reinforce the narratives of individual genius that historians of science have long tried to overturn.

Narratives of individual greatness are also kept alive and intact by political and economic pressures facing the world of scholarship. Universities, increasingly driven by market concerns, invest heavily in the promotion and marketing of their most successful employees, hoping in this way to attract students as well as gifts from high-profile donors. Academics compete not only for research funding but also for the attention of agents and publishers, driven by the sincere desire to communicate their research to broader audiences but also by ever-changing expectations of what it means to be a successful academic. And while trust in science remains high even as trust levels in global leaders and journalists fall, the public is showing increasing concerns about governmental influence on scientific research.[5]

It is in this light that we should understand the most recent recalibration of Galileo as a public figure, one that entails his appropriation by groups keen to incite and exploit doubts about the value of scientific research; various divisive figures who deny humanity's impact on climate change, vaccine safety, or even that the earth is round have now embraced Galileo as their hero and shining example. The move even has a name: In pulling a so-called Galileo Gambit, they assert that they bravely confront and resist the false majority opinion out of an admirable devotion to reason and that the resistance they immediately experience means that they will later be proven right.[6] This newest configuration of Galileo's fame repurposes the idea of him as an antiauthoritarian figure, ready to upset traditional order, to now shed doubt on scientific findings.

This newest reinterpretation of Galileo's story may be particularly jolting, but it also underscores that the creation, dissemination, and consolidation of individual narratives of greatness remains an inherently collective process, as well as one that exposes wider fissures in society. Groups within and outside the narrow world of learning continue to look to famous individuals as a way to articulate their hopes, desires, expectations, and fears for the future. Those hopes and expectations often clash, leading to confrontations and conflicts. This book has shown that this is nothing new and that our own ongoing discussions about the value, reach, and impact of scientific research and, above all, about the proper ways to bestow recognition for scholarly contributions echo the way Galileo's contemporaries debated the value and dangers of his discoveries and his fame.

I have traced their discussions and their construction of Galileo's fame not with the intention to diminish or dismiss the quality of his achievements but to illuminate the careful attention that went into cultivating public appreciation for his findings, as well as to show that even narratives of individual greatness are in fact the result of collective efforts. This means that these narratives are forever subject to change, and in this sense even the posthumous glory Galileo's contemporaries praised above all is neither stable nor constant. Scholarly fame rarely remains the same for long.

ACKNOWLEDGMENTS

MOST FIRST MONOGRAPHS HAVE THEIR ORIGINS IN A doctoral thesis. This one knows an even longer history: My interest in Galileo ignited in 2014, when I followed a master's seminar on science and religion in Rome, and it has followed me around the world ever since. This book bears the marks of the myriad institutions of which I was lucky enough to be a part, as well as of numerous conversations I had over the years with wonderful colleagues and friends.

My deepest debt I owe to my doctoral advisor, Arnoud Visser, who introduced me to the topic of fame as a theme of academic study in one of his seminars at Utrecht University. Arnoud supported the project from its inception, read draft after draft, and was the kindest, most generous, and most encouraging mentor I could have asked for. When I took my first steps into the history of science, I benefited from Floris Cohen's enthusiasm and critical eye; I also thank him for introducing me to the larger *Isis* family. The Dutch Research Council (NWO) fully funded my doctoral research, for which I am grateful. In 2018, I spent a few glorious months at the Medici Archive Project in Florence, where I benefited enormously from the wisdom and guidance of its members; my thanks similarly go out to the staff at the Museo Galileo. For their thoughtful questions and advice at a crucial point in the project, I thank Sven Dupré, Harald Hendrix, Inger Leemans, Maartje van Gelder, and Frans van Lunteren. I am especially grateful to Maartje van Gelder for her extraordinary mentorship and exquisite book recommendations before, during, and after my PhD. I hope I get the opportunity to pay it forward.

A fellowship from the Niels Stensen Foundation allowed me to spend most of the academic year of 2021–2022 in Los Angeles to work on my book manuscript. This was truly an unforgettable experience, especially after spending so much of the previous year holed up in dreary, dark Amsterdam. I am particularly grateful to Jake Soll, who read numerous drafts, offered indispensable writing advice, and sustained me and Arthur with stimulating conversation, wine, and the best couscous I have tasted to date. I was fortunate enough to discuss with Mario Biagioli all things Galileo, the history of Echo Park, and the joy of driving in LA. He will be sorely missed. My thanks also go out to Nathan Perl-Rosenthal and Jessica Marglin for welcoming us into their home during the first weeks of our stay.

Subsequently, I was lucky enough to find not one but two marvelous temporary homes at Cambridge University. I benefited tremendously from the numerous seminars and working groups at the Department of History and Philosophy of Science and especially from the kindness and generosity of Dániel Margócsy, who commented on various drafts and provided advice and encouragement during the publication process. On the other side of town, Jesus College provided a quiet working space, a wonderfully interdisciplinary academic community, a brilliant bread pudding featuring croissants, and the exciting opportunity to spot fox cubs and deer on its expansive grounds. A brief but fabulous return to California in 2024 was made possible by an invitation from Hannah Marcus and Paula Findlen, who have supported my work in various ways since 2017 and welcomed me into the world of Italo-American *Galileisti*, for which I am deeply grateful. The KNIR–Van Woudenberg Dissertation Prize enabled me to return to Rome almost ten years after taking that first seminar and to discuss scholarly fame through the ages in the serene environment of the Royal Netherlands Institute. I thank all workshop participants for their comments and suggestions.

At the University of Pittsburgh Press, Abby McAllister has patiently worked with me on this book, and I am grateful for her enthusiasm and support. I am also indebted to the two anonymous reviewers who read the entire manuscript, asked probing questions, corrected mistakes, and came up with detailed and welcome suggestions that greatly improved the book's final form. The book's final form was also much improved by Maureen Bemko, who meticulously copyedited my prose, while Kelly Lynn Thomas and Alex Mathews expertly guided the book through the production process. Thank you to both. Any mistakes remaining are of course entirely my own.

Finally, I want to thank the family and friends, and Sam and Josephine in particular, who provided me with kind words, encouragement, and distractions throughout the years. My parents, Mirjam and Michiel, have little patience for the past and suggested I do something "useful, like change management." They nonetheless supported me in every way possible when I instead turned to history. They are the kindest, most loving people I know, and I am deeply grateful for everything they have taught me. For over a decade now, Arthur has accompanied me on most of my Galileo and non-Galileo-related travels around the world, enthusiastically tagging along to Bologna, Padua, Florence, New Orleans, and Los Angeles—to name just a few places—and only slightly less enthusiastically to colder, less sunny environments. After all these peregrinations, it is such a joy to be back home—in Amsterdam, with Freddy.

NOTES

INTRODUCTION

1. Henry Wotton to Robert Cecil, 13 March 1610, in Galileo Galilei, *Le opere di Galileo Galilei: Appendice*, vol. 2, *Carteggio*, ed. Michele Camerota and Patrizia Ruffo, in collaboration with Massimo Bucciantini (Giunti Editore, 2015), 47. This book, like all Galileian scholarship, is heavily indebted to Antonio Favaro's edition of Galileo's collected works: *Le opere di Galileo Galilei*, 20 vols. (Barbèra, 1890–1909; reprint, 1929–1939). After the initial citation of a volume of these collected works, I then cite them in abbreviated form as *GO* plus the volume and page numbers (e.g., *GO* app. 2.47, for the Wotton letter cited above).

2. Galileo Galilei, *Le opere di Galileo Galilei: Appendice*, vol. 1, *Iconografia Galileiana*, ed. Federico Tognoni (Giunti Editore, 2013), 52–53.

3. For the LEGO figures, see "LEGO Figures Aboard Juno," Jet Propulsion Laboratory, 3 August 2011, https://www.jpl.nasa.gov/images/pia14413-lego-figurines-aboard-juno/.

4. The LEGO figures have gotten attention from several international news outlets; see, for instance, Sarah Kaplan, "Why a Tiny Lego Version of Galileo Rode on NASA's Juno Probe All the Way to Jupiter," *Washington Post*, 5 July 2016, https://www.washingtonpost.com/news/speaking-of-science/wp/2016/07/05/why-a-tiny-lego-version-of-galileo-rode-on-nasas-juno-probe-all-the-way-to-jupiter/.

5. The telescope was invented in 1608, perhaps by the Dutch glassmaker Hans Lipperhey, who first applied for a patent for the invention. It was initially used as a military instrument; Galileo came to hear of it in the summer of 1609 and started work on improving it, but during this same summer the Englishman Thomas Harriot had used the telescope to observe the surface of the moon. Galileo aimed his instrument toward the sky in the autumn of 1609, and on the evening of 7 January 1610 he observed what he later realized were satellites orbiting Jupiter. Diligent-

ly keeping notes over the next few weeks and working with remarkable speed, Galileo managed to publish his findings at the end of March, claiming that he had been the first to discover these previously unknown celestial bodies and asserting the right to name them after the patrons he wished to honor. When the German mathematician and astronomer Simon Marius (or Simon Mayr) later argued that he had observed the satellites days or even weeks before Galileo had detected them through his telescope, Galileo's reputation as their discoverer had already been consolidated to such an extent that Marius's claims stood no chance of success. Marius used the Julian calendar, whose dates ran ten days behind the Gregorian one; his 29 December thus equals 8 January in the Gregorian calendar Galileo used. On Harriot, see Stephen Pumfrey, "Harriot's Maps of the Moon: New Interpretations," *Notes and Records of the Royal Society of London* 63, no. 2 (2009): 163. On Marius, see Jay M. Pasachoff, "Simon Marius's *Mundus Iovialis*: 400th Anniversary in Galileo's Shadow," *Journal for the History of Astronomy* 46, no. 2 (2015): 220–21.

6. Zorzi to the Doge and Senate in Venice, 27 November 1627, in *Calendar of State Papers Relating to English Affairs in the Archives of Venice*, vol. 20, 1626–1628, ed. Allen B. Hinds (His Majesty's Stationery Office, 1914), 478–96. I am grateful to Nina Lamal for sharing this reference with me. For other examples, see a letter by Leonardo Accolti to Giovanni Battista Bartolini Baldelli, 1 June 1614, Mediceo del Principato, vol. 1700, via the Medici Archive Project, and an anonymous poem celebrating a list of beautiful women and claiming it is impossible to find fault with them, even if one were to use Galileo's telescope, in Archivio di Stato di Firenze (ASF), Mediceo del Principato, vol. 6424. Unless otherwise stated, all translations are mine.

7. On fama's personifications, see Philip Hardie, *Rumour and Renown: Representations of Fama in Western Literature* (Cambridge University Press, 2012); and Gianni Guastella, *Word of Mouth: Fama and Its Personifications in Art and Literature from Ancient Rome to the Middle Ages* (Oxford University Press, 2017), 5.

8. The most captivating example is probably that of Herostratus, who in 356 BC burned down Artemis's temple in Efeze to gain eternal fame. Arnoud Visser, *In de gloria: Literaire roem in de Renaissance* (Inaugural Lecture, Utrecht University, printed in The Hague, 2013), 5; and Leo Braudy, *The Frenzy of Renown: Fame and Its History* (Oxford University Press, 1986), 51.

9. For the shift from ancient Roman ideals of renown to early Christian ones, see Braudy, *Frenzy of Renown*, 115–17, 150–89. For a concise discussion on how the Christian suspicion of earthly renown drew on ancient philosophy, see Hardie, *Rumour and Renown*, 22–36.

10. For a discussion of fama's etymology, see Hardie, *Rumour and Renown*, 1–3; on rumor, see Keith M. Botelho, *Renaissance Earwitnesses: Rumor and Early Modern Masculinity* (Palgrave Macmillan, 2009); and Claire Walker and Heather Kerr, eds., *Fama and Her Sisters: Gossip and Rumour in Early Modern Europe* (Brepols, 2015).

11. Frances Gage, "Caravaggio's *Rumore*: Fact, Fiction and Authority in Giovanni Baglione's *Lives of the Painters, Sculptors and Architects*," *Past & Present* 257, supplement 16 (2022): 126; Hardie, *Rumour and Renown*, 2–3, 22–26; Guastella, *Word of Mouth*, 5.

12. On the many other duplicities of fama, see Hardie, *Rumour and Renown*, 2–11.

13. Luca Valerio to Galileo, 4 April 1609, *GO* 10.240: "Fama è di due sorti: l'una, figlia del volgo, nata per forza de' suoi stolidi gridi, la quale V. S. con ragione disprezza; l'altra è quella che nasce da pochi huomini et savi, che con la loro autorità et signoria naturale piegano et volgono a segno ragionevole lo sfrenato giuditio della plebe: et questa fama è stabile et degna del nome; l'altra, a guisa d'animale imperfetto, sorto dalle brutture della materia, oltraggiata dal tempo, a pena nata muore."

14. Thomas Kuehn, "*Fama* as a Legal Status in Renaissance Florence," in *Fama: The Politics of Talk and Reputation in Medieval Europe*, ed. Thelma Fenster and Daniel Lord Smail (Cornell University Press, 2003); and Chris Wickham, "*Fama* and the Law in Twelfth-Century Tuscany," in *Fama*, ed. Fenster and Smail; Francesco Migliorino, *Fama e infamia: Problemi della società medievale nel pensiero giuridico nei secoli XII e XIII* (Editrice Giannotta, 1985).

15. On reputation, see Kerr and Walker, *Fama and Her Sisters*; and Una McIlvenna, *Scandal and Reputation at the Court of Catherine de Medici* (Routledge, 2016).

16. Besides Kuehn, "*Fama* as a Legal Status in Renaissance Florence," and Wickham, "*Fama* and the Law in Twelfth-Century Tuscany," also see Elisabeth Horodowich, "The Gossiping Tongue: Oral Networks, Public Life and Political Culture in Early Modern Venice," *Renaissance Studies* 19, no. 1 (2005): 22–45; and Elisabeth Horodowich, "The Meanings of Gossip in Sixteenth Century Venice," in *Spoken Word and Social Practice: Orality in Europe (1400–1700)*, ed. Thomas V. Cohen and Lesley K. Twomey (Brill, 2015).

17. Francis Bacon, *The Essays or Counsels, Civil and Moral*, ed. and intro. Brian Vickers (Folio Society, 2002), 207–8.

18. Botelho, *Renaissance Earwitnesses*, 1–8.

19. Bacon, *Essays or Counsels, Civil and Moral*, 207.

20. Bacon, *Essays or Counsels, Civil and Moral*, 7.

21. Thelma Fenster and Daniel Lord Smail, eds., introduction to *Fama: The Politics of Talk and Reputation in Medieval Europe* (Cornell University Press, 2003), 5–8; Clare Haru Crowston, *Credit, Fashion, Sex: Economies of Regard in Old Regime France* (Duke University Press, 2013), 3–6.

22. Patricia Fara, *Newton: The Making of Genius* (Columbia University Press, 2002). On the nineteenth-century development of Newton's image, see Rebekah Higgitt, *Recreating Newton: Biographies of Newton and the Making of Nineteenth-Century History of Science* (Pickering & Chatto, 2007); Andrew Pettegree, *Brand Luther: 1517, Printing, and the Making of the Reformation* (Penguin Random House 2015); and Peter Burke, *The Fabrication of Louis XIV* (Yale University Press, 1992).

23. Mario Biagioli, *Galileo Courtier: The Practice of Science in the Culture of Absolutism* (University of Chicago Press, 1994). The pathbreaking work on self-fashioning is of course Stephen Greenblatt, *Renaissance Self-Fashioning: From More to Shakespeare* (University of Chicago Press, 1980). A good entry point into the literature on early modern brokers is Marika Keblusek and Badeloch Vera Noldus, eds., *Double Agents: Cultural and Political Brokerage in Early Modern Europe* (Brill, 2011). Janie Cole specifically discusses reputation brokering in "Cultural Clientelism and Brokerage Networks in Early Modern Florence and Rome: New Correspondence Between the Barberini and Michelangelo Buonarroti the Younger," *Renaissance Quarterly* 60, no. 3 (2007): 729–88.

24. Massimo Bucciantini, Michele Camerota, and Franco Giudice, *Galileo's Telescope: A European Story*, trans. Catherine Bolton (Harvard University Press, 2015); Renée Raphael, *Reading Galileo: Scribal Technologies and the "Two New Sciences"* (Johns Hopkins University Press, 2017); Renée Raphael, "Making Sense of Day 1 of the *Two New Sciences*: Galileo's Aristotelian Inspired Agenda and His Jesuit Readers," *Studies in History and Philosophy of Science* 42, no. 4 (2011): 479–91; Renée Raphael, "Reading Galileo's *Discorsi* in the Early Modern University," *Renaissance Quarterly* 68, no. 2 (2015): 558–96; Renée Raphael, "Galileo's *Two New Sciences* as a Model of Reading Practices," *Journal of the History of Ideas* 77, no. 4 (2016): 539–65.

25. Maurice A. Finocchiaro, *Retrying Galileo, 1633–1992* (University of California Press, 2005); Michael Segre, *In the Wake of Galileo* (Rutgers University Press, 1991); Stefano Gattei, ed. and trans., *On the Life of Galileo: Viviani's Historical Account and Other Early Biographies* (Princeton University Press, 2019); Paula Findlen, "Long After the Trial: Galileo's Rediscovery, Florentine Nostalgia, and Enlightened Passions," in *Florence After the Medici*, ed. Corey Tazzara, Paula Findlen, and Jacob Soll (Routledge, 2019).

26. Robert van Krieken, "Celebrity's Histories," in *Routledge Handbook of Celebrity Studies*, ed. Anthony Elliott (Routledge, 2018).

27. Antoine Lilti, *The Invention of Celebrity: 1750–1850*, trans. Lynn Jeffress (Polity, 2017), 9. Other useful entry points into the vast field of celebrity studies include Fred Inglis, *A Short History of Celebrity from Byron to Beckham* (Princeton University Press, 2010); and Robert van Krieken, *Celebrity Society* (Routledge, 2012). Leo Braudy's *Frenzy of Renown* deals more broadly with historical fame through the ages.

28. Lilti, *Invention of Celebrity*, 8–10.

29. Van Krieken, *Celebrity Society*, 5–6.

30. The gift economy and the market are of course not strictly separate but bled into one another; see Natalie Zemon-Davis, *The Gift in Sixteenth-Century France* (Oxford University Press, 2000); Natalie Zemon-Davis, "Beyond the Market: Books as Gifts in Sixteenth-Century France," *Transactions of the Royal Historical Society* 33 (1983): 69–88; Avner Offer, "Between the Gift and the Market: The Economy of Regard," *Economic History Review* 50, no. 3 (1997): 450–76; Laurence Fontaine, *The Moral Economy: Poverty, Credit, and Trust in Early Modern Europe* (Cambridge University Press, 2014); Crowston, *Credit, Fashion, Sex*; and Dániel Margócsy, *Commercial Visions: Science, Trade, and Visual Culture in the Dutch Golden Age* (University of Chicago Press, 2014).

31. Emilio Costa, "Galileo e lo Studio di Bologna," *Studi e Memorie per la Storia dell'Università di Bologna* 7 (1992): 8–11. I discuss the episode in more detail in the section titled "The Limits of Reputation" in chapter 1 of this book.

32. *GO* 19.487–90, esp. 488: "Secondariamente, conduconsi lettori delle scienze et arti non solamente per la particulare utilità degli scolari privati che a quelle attendono, ma ancora per reputazione et honorevolezza di esse Università si cerca di havere i più insigni e famosi professori di quelle. Non si dubita punto che il Sig.r Galileo si sia talmente avanzato di nome e fama in queste scienze, che forse nessun altro all'età nostra gli metta il piede innanzi." Also *GO* 19.489: "Che quel danaro serva in utile e servizio di quello Studio: del quale che maggior servizio può esser di quello onde gli viene splendore e reputazione? E se ciò gli venga apportato dal S.r Galileo, lascio giudicare a chi sa gli honori che egli ha riceuti e riceve da' primi principi del mondo e da tutti i letterati famosi di Europa, che l'hanno celebrato con i loro scritti, la cui gloria nessuno mi negherà che non redondi in illustrazione dello Studio di Pisa, poi che il Sig.r Galileo si intitola suo primario Matematico."

33. The various strands of debate regarding early modern credibility are excellently discussed in R. W. Serjeantson, "Proof and Persuasion,"

in *The Cambridge History of Science*, vol. 3, *Early Modern Science*, ed. Katherine Park and Lorraine Daston (Cambridge University Press, 2006).

34. Steven Shapin and Simon Schaffer, *Leviathan and the Air-Pump: Hobbes, Boyle, and the Experimental Life* (Princeton University Press, 1985); Steven Shapin, *A Social History of Truth: Civility and Science in Seventeenth-Century England* (University of Chicago Press, 1994). For a useful overview of early responses to Shapin and Schaffer's *Leviathan and the Air-Pump*, see Azadeh Achbari, "The Reviews of *Leviathan and the Air-Pump*: A Survey," *Isis* 108, no. 1 (2017): 108–16.

35. Biagioli, *Galileo Courtier.*

36. On the ideals and practices of the Republic of Letters, see Hans Bots and Françoise Waquet, *La République des Lettres* (De Boeck, 1997); Anthony Grafton, "A Sketch Map of a Lost Continent: The Republic of Letters," *Republics of Letters: A Journal for the Study of Knowledge, Politics, and the Arts* 1 (1 May 2009): 1–18; Anne Goldgar, *Impolite Learning: Conduct and Community in the Republic of Letters, 1680–1750* (Yale University Press, 1995); and Lorraine Daston, "The Ideal and Reality of the Republic of Letters in the Enlightenment," *Science in Context* 4, no. 2 (1991): 367–86.

37. Margócsy, *Commercial Visions*, 8–9, 18–19, 24.

38. Inger Leemans and Anne Goldgar, eds., *Early Modern Knowledge Societies as Affective Economies* (Routledge, 2020), 1–32.

39. Barbara Shapiro, *A Culture of Fact: England, 1550–1720* (Cornell University Press, 2000), 13–18.

40. Fenster and Smail, *Fama*; Horodowich, "Meanings of Gossip."

41. Andrea Frisch, *The Invention of the Eyewitness: Witnessing and Testimony in Early Modern France* (University of North Carolina Press, 2004), 11–19.

42. Lilti, *Invention of Celebrity.*

43. John Heilbron, *Galileo* (Oxford University Press, 2010); Michele Camerota, *Galileo Galilei e la cultura scientifica nell'età della Controriforma* (Salerno Editrice, 2004).

1. Before Fame

1. For works that pay explicit attention to this period, see Heilbron, *Galileo*; and Camerota, *Galileo Galilei e la cultura scientifica*.

2. On reputation in the courtroom, see Daniel Lord Smail, *The Consumption of Justice: Emotions, Publicity, and Legal Culture in Marseille, 1264–1423* (Cornell University Press, 2013); and Fenster and Smail, *Fama*. Steve Shapin and Simon Schaffer have famously stressed the importance of noble status and subscription to gentlemanly norms; see Shap-

in and Schaffer, *Leviathan and the Air-Pump*. Andrea Frisch, in *Invention of the Eyewitness*, has argued that in France experience and expertise replaced social standing and reputation over the course of the sixteenth century, but Barbara Shapiro's *Culture of Fact*, 13–18, demonstrates the continued relevance of reputation and rumor in seventeenth-century English courtrooms.

3. Among those scholarly achievements this chapter does not address are his lectures for the Accademia Fiorentina in 1588, for which see Heilbron, *Galileo*, 28–33; and his work with the lodestone in 1608, on which see Richard S. Westfall, "Science and Patronage: Galileo and the Telescope," *Isis* 76, no. 1 (1985): 11–30; and Biagioli, *Galileo Courtier*, 120–26.

4. *GO* 19.47. The guild convened in the Palazzo dell'Arte dei Giudicai e Notai in the Via del Proconsolo until the sixteenth century.

5. Maddalena Ricasoli and Jacopo Quaratesi filed their case with the Signori Luogotenenti e Consiglieri della Repubblica Fiorentina, who filled in for the Magistrato Supremo, the city's highest executive organ. *GO* 19.4–45; Elisa Goudriaan, *Florentine Patricians and Their Networks: Structures Behind the Cultural Success and the Political Representation of the Medici Court (1600–1660)* (Brill, 2018), 9. On the Florentine legal system and changes therein over the course of the fourteenth to sixteenth centuries, see Lauro Martines, *Lawyers and Statecraft in Renaissance Florence* (Princeton University Press, 2016).

6. Smail, *Consumption of Justice*, 51–52; Kuehn, "*Fama* as a Legal Status in Renaissance Florence."

7. *GO* 19.60: "Item, che cosa sia publica voce et fama, et quante persone la faccino, et che circumstantie vi si ricerchino di ragione." Galileo replied, "Al 38a rispose, la publica voce et fama essere quando ogn'uno, o la maggior parte, concorre nel medesimo dire, et nel resto rimettersi alle leggi." This questioning of Galileo was in line with broader legal practices. Smail, *Consumption of Justice*, 58.

8. Horodowich, "Gossiping Tongue," 42–43; Wickham, "*Fama* and the Law," 17–23. For an entertaining example of one interrogator aiming to establish precisely how many people made up the publica fama, see Wickham, "*Fama* and the Law," 17; the answers he received ranged from twenty to the frustratingly vague "a majority of 'the locality.'"

9. *GO* 19.60: "Item, se può essere che una cosa si dica publicamente et notoriamente, et che sia publica et notoria, et che di poi si trovi non esser vera." And Galileo's reply: "Al 37a rispose, poter essere che una cosa si dica per vera publicamente, et poi non sia stata vera." And, later: "Item, se può essere che uno per publica voce et fama sia reputato in un modo, et che nel vero la cosa stia altrimenti." And, at *GO* 19.84: "Al 17a rispo-

se, esser quasi che impossibile che una publica voce et fama, continuata molto tempo, sia falsa."

10. *GO* 19.84.

11. *GO* 19.84: "Item, quali actioni sieno state vedute et raccontate da persone degne di fede, che potessino fare detta publica voce et fama, et quali sieno state quelle persone degne di fede, et se erono gentilhuomini o di che qualità persone, et se maggiori d'ogni exceptione."

12. *GO* 19.84: "Al 14e rispose, che le attioni che hanno fatto reputare detto Giovambatista oppresso da humori malenconici sono state vedute et raccontate da più persone degne di fede, et in particolare da Mess. Lorenzo Giacomini, Mess. Francesco Guadagni, Mess. Iacopo Quaratesi, Mess. Neri Ricasoli et Giovanni suo fratello, Mess. Giovambattista Strozzi, Mess. Bernardo de'Bardi, Mess. Giulio da Barga fisico, Mess. Agnolo Bonelli fisico, un medico che lo medicò in Genova, un Teatino di Genova, et da altri che di presente non si ricorda." The adjective "Teatino" refers to inhabitants of Chieti, a town in the Abruzzo region in Italy, as well as more generally to people from that region.

13. *GO* 19.47–108, esp. 55–60 and 72–76. The importance assigned to Giambattista's clothing corresponds to recent scholarship on early modern apparel; see especially Ulinka Rublack, *Dressing Up: Cultural Identity in Renaissance Europe* (Oxford University Press, 2010); and Fontaine, *Moral Economy*, 268–75.

14. For the question on Naples and Rome, see *GO* 19.72–76.

15. For various studies across Europe, see Shapiro, *Culture of Fact*, 12–14; Kuehn, "*Fama* as a Legal Status," 30; and Jeffrey A. Bowman, "Infamy and Proof in Medieval Spain," in *Fama: The Politics of Talk and Reputation in Medieval Europe*, ed. Thelma Fenster and Daniel Lord Smail (Cornell University Press, 2003), 96.

16. *GO* 19.63: "Item, se s'ha da credere più a 4 gentil'huomini, che a servitori, vetturini, hosti o garzoni." Galileo's answer is on the same page: "Al 59a rispose, che per lo pari, cioè in cose che possono essere note a tutte le gente di che nell'interrogatorio, si deve credere più a gentil'huomini che agl'altri."

17. On challenges to the standing of witnesses, or "exceptions," more broadly, see Smail, *Consumption of Justice*, 25, 60.

18. *GO* 19.46: "Galileo, sfratato, figliuolo d'un maestro di sonare"; "Galileo. Fu frate monaco di Valombrosa, figliuolo d'un maestro di sonare di liuto"; "figliuolo d'un sonatore di liuto, povero e sfratato, *ecc.*"

19. *GO* 19.46: "sfratato"; "Galileo sfratato"; "o fu uno frate sfratato senza discretione."

20. Heilbron, *Galileo*, 3–4; Camerota, *Galileo Galilei*, 30–42.

21. Later on, after Galileo's breakthrough, the Vallombrosan order

was proud to have once included Galileo among its ranks; a monk writing the history of illustrious men of the order wrote, "We should not omit the celebrated name of Galileo Galilei, great mathematician. He was a Vallombrosan novice, and took the first exercises of his admirable *ingegno* at the Vallombrosan school." *GO* 19.46, 53.

22. On the theft, see *GO* 19.85, 87. On the bribery, see *GO* 19.46: "Tristaccio, scortese, sfratato! et poi perché t'hanno promesso fior. 150 per la sorella, far questo etc. falsamente!" and *GO* 19.45: "Queste cose non le conta alcuno, se non questo per l'anima del far sua sorella monacha."

23. *GO* 19.87–88.

24. Some months later, however, the couple initiated a second trial, in which they fought the validity of several bequests they were supposed to make according to yet another testament left behind by Giambattista. *GO* 19.44–45; Heilbron, *Galileo*, 23–27.

25. In this section and the next, I will refer to the Bolognese, Paduan, Pisan, and Florentine institutions of higher learning occasionally as "universities" but mostly with the term *studios*, which seems to have been preferred by the historical agents under discussion.

26. Antonio Favaro, "Galileo ed il Magini: Aspiranti ad una lettura di matematica nello Studio di Bologna," *Atti del R. Istituto veneto di scienze, lettere ed arti* 82 (1922–1923): 148–49.

27. Enrico Caetani to the Bolognese senate, 10 February 1588, *GO* 20.26: "Tengono pensiero le SS.rie VV., per quello che mi s'espone di condurre un matematico alla lettura pubblica dello Studio di Bologna, et intendo che sia stato loro proposto M. Galileo Galilei, nobile fiorentino, il quale habbia grande approbatione della sua sufficienza. Se le SS.rie VV. inclinaranno a condurlo, aggiongo la mia raccomandatione a beneficio suo, acciò nel concorso delli altri li giovi appresso la loro humanità l'esser raccomandato da me, che ne sentirà particolar obligo alle SS.rie VV.; alle quali mi offero con tutto l'animo. Di Roma, a' 10 di Febbraio 1588. Delle SS. VV. molto Ill.ri. Come fratello per servirle. Il Cardinale Gaetano. Alli SS.ri Quaranta del Reggimento di Bologna."

28. Paul F. Grendler, *The Universities of the Italian Renaissance* (Johns Hopkins University Press, 2002), 14–18; David Lines, "Papal Power and University Control in Early Modern Italy: Bologna and Gregory XIII," *Sixteenth Century Journal* 43, no. 3 (2013): 663–82.

29. *GO* 19.39; Carlo Malagola, "Galileo Galilei e L'università di Bologna: Memoria del Dott. Carlo Malagola," *Archivio Storico Italiano*, 4th ser., 7, no. 122 (1881): 190–91. On this practice of witnessing and recommendation, see also Margaret E. Schotte, *Sailing School: Navigating Science and Skill, 1550–1800* (Johns Hopkins University Press, 2019), 18.

30. *GO* 19.36: "Mathematico Fiorentino raccomandato dal S.r Ar-

tani. M. Galileo Galilei, nobile Fiorentino, giovane d'anni 26 incirca, è istruttissimo in tutte le scienze matematiche, ed è allievo di M. Ostilio Ricci, huomo segnalatissimo e provvisionato dal Gran Duca Francesco di felice memoria, del quale ci sono anco fedi in commendazione del valor di questo giovane. Fu condotto alla lettura pubblica di Matematica in Siena; s'è esercitato assai privatamente, ed ha letto a molti gentiluomini e in Firenze e in Siena. È di grandissimo giudicio in questo e in molte altre cose nelle quali ha posto studio, come in particolare nell'Umanità e nella Filosofia e in altre belle qualità. Al presente domanda e desidera la lettura di Matematica in questa città, offerendosi prontamente a concorrere nel merito con qual si voglia altro di questa professione, in qualunque modo bisognerà."

31. Malagola, "Galileo Galilei e L'università di Bologna," 191; Heilbron, *Galileo*, 8, 28–33.

32. Grendler, *Universities of the Italian Renaissance*, 160.

33. The witnesses—Giovanni Bardi de'Conti di Vernio, Giovanni Battista Strozzi, Luigi di Piero Alamanni, and Giambattista Ricasoli—testified on 12 December 1587. *GO* 1:183–84. On the theorems, see Matteo Valleriani, *Galileo Engineer* (Springer, 2010), 15–17; Heilbron, *Galileo*, 34–41; Favaro, "Galileo ed il Magini"; and Camerota, *Galileo Galilei*, 46–50.

34. *GO* 1.183–84: "Io Gioseppe Moleto, Lettor publico delle Mathematiche nello Studio di Padova, dico haver letto i presenti Lemma et Theorema, i quali mi son parsi buoni, e stimo l'autor d'essi esser buono et esercitato Geometra. Il medesimo Gioseppe ha scritto di Man propria." Galileo's correspondence includes a further approval of the theorems, though it is unfortunately undated and anonymous: "Il mio amico loda infinitamente lo inventore di questa speculatione, et insieme col Sig.r Moleto lo giudica molto versato nelle matematiche." This praise is followed by another opinion, possibly from someone in Bologna, that is much more critical; see *GO* 10.21–22.

35. Cited in Costa, "Galileo e lo Studio di Bologna," 10–11: "È vero che altre volte si è havuto in pensiero di condurre in questo Studio alla lettura di Mattematica persona di nome et fama che de gli ordinari ne siamo forniti; ma da un pezzo in qua non se n'essendo parlato, non potiamo rispondere alla lettera di V.S. Ill.ma per la quale si raccomanda m. Galileo Galilei nobile fiorentino, se non caso che torni in piedi tal maneggio, prometterle come facciamo di haver in memoria la racommandatione di lei et le buone qualità di lui, con desiderare sempre in tutto quello che per noi si potrà di servire e sodisfare a. V.S. Ill.ma et R.ma alla quale reverenti basciamo la mano."

36. On the chair of mathematics in Bologna, see Alexander Marr,

Between Raphael and Galileo: Mutio Oddi and the Mathematical Culture of Late Renaissance Italy (University of Chicago Press, 2011), 64; and Grendler, *Universities of the Italian Renaissance*, 416–28.

37. "È vero che altre volte si è havuto in pensiero di condurre in questo Studio alla lettura di Mattematica persona di nome et fama che de gli ordinari ne siamo forniti." Quoted in Costa, "Galileo e lo Studio di Bologna," 10.

38. Grendler, *Universities of the Italian Renaissance*, 159–60.

39. On Magini, see Favaro, "Galileo ed il Magini," 150–51; and Robert S. Westman, *The Copernican Question: Prognostication, Skepticism, and Celestial Order* (University of California Press, 2011), 435. Mario Biagioli, *Galileo's Instruments of Credit: Telescopes, Images, Secrecy* (University of Chicago Press, 2006), 5, points out that mathematicians—because of their limited audience—published relatively little and preferred to communicate their findings through letters. However, this seems to apply mostly to mathematicians working specifically on theoretical mathematics, as those working on astronomy and astrology generally published more.

40. Guidobaldo del Monte to Galileo, 30 December 1588, *GO* 10.39: "Ho anche con grandissima mia satisfattione sentito ch'ella vogli mandar fuori le sue cose del centro della gravezza, che in verità V. S. ne acquistarà molto honore."

41. Westman, *Copernican Question*, 207; Heilbron, *Galileo*, 40–41; William R. Shea and Mariano Artigas, *Galileo in Rome: The Rise and Fall of a Troublesome Genius* (Oxford University Press, 2003), 1–15.

42. Michel Coignet to Galileo, 31 March 1588, *GO* 10.31; Stillman Drake, *Galileo at Work: His Scientific Biography* (University of Chicago Press, 1978), 12.

43. Biagioli, *Galileo Courtier*, 24–31; Grendler, *Universities of the Italian Renaissance*, 28–29, 73–75.

44. *GO* 10.47–49.

45. Heilbron, *Galileo*, 84.

46. Valleriani, *Galileo Engineer*, 117–54.

47. *GO* 2.278: "Testes vos estis, numerosa iuventus, qui huc convolastis, ut me de hac admiranda apparitione disserentem audiatis." See also David Freedberg, *The Eye of the Lynx: Galileo, His Friends, and the Beginnings of Modern Natural History* (University of Chicago Press 2002), 84–86; and Alessandro De Angelis, "Galileo Galilei and a Forgotten Poem on the 1604 Supernova," *Quaderni di Storia della Fisica* 1, no. 1 (2023): 17–29.

48. Freedberg, *Eye of the Lynx*, 86–90; Edward Muir, *The Culture Wars of the Late Renaissance: Skeptics, Libertines, and Opera* (Harvard University Press, 2007), 33–40, 50–51.

49. On Galileo's teaching in this period, see Biagioli, *Galileo's Instruments of Credit*, 7. For the correspondence of Kepler and Galileo, see *GO* 10.69–71; and Westman, *Copernican Question*, 15–16, 357–66. For Brahe's attempts to get in contact with Galileo, see *GO* 10.78–79.

50. Aldo Stella, "Galileo, il circolo culturale di Gian Vincenzo Pinelli e la 'Patavina libertas,'" in *Galileo e la cultura padovana: Convegno di studio promosso dall'Accademia Patavina di Science Lettere ed Arti nell'ambito delle celebrazioni galileiane dell'Università di Padova 13–15 febbraio 1992*, ed. Giovanni Santinello (CEDAM, 1992); Nick Wilding, *Galileo's Idol: Gianfrancesco Sagredo and the Politics of Knowledge* (University of Chicago Press, 2014); Gaetano Cozzi, *Paolo Sarpi tra Venezia e l'Europa* (Einaudi, 1979).

51. Tognoni, *GO* app. 1.27–28. Recently, the art historian Vincenzo Mancini proposed that the portrait was actually painted by Francesco Apollodoro, but this is rejected by Tognoni.

52. Galileo Galilei, *Difesa di Galileo Galilei contro alle calunnie & imposture di Baldassar Capra Milanese* (Venice, 1607). I here rely on and refer to the edition in *GO* 2.513–601 (quote on 517).

53. Galileo's conflict with Capra has most fruitfully been discussed in Biagioli, *Galileo's Instruments of Credit*, 7–13; Camerota, *Galileo Galilei*, chap. 3, sec. 5; and Wilding, *Galileo's Idol*, 43–49. My analysis here is especially indebted to Wilding's work, which first points to the use of witnessing to create an authoritative network, both in the courtroom and in print.

54. Galileo Galilei, *Le Operazioni del compasso geometrico e militare* (Padua, 1606). The work is reproduced in *GO* 2.363–424; here and in my discussion of Capra's work, I refer to these reproductions.

55. Baldassare Capra, *Usus et fabrica circini cuiusdam proportionis, opera et studio* (Padua, 1607), reproduced in *GO* 2.425–511.

56. Biagioli, *Galileo's Instruments of Credit*, 9; Wilding, *Galileo's Idol*, 44.

57. Galileo, *Difesa*, 536–37. Quote on 537: "e non senza sdegnose esclamazioni mi fece vedere la insolenza usata dal Capra."

58. Galileo to Riformatori dello Studio, 9 April 1607, *GO* 10.172: "usurparsi l'opera et l'honore mio" and "provegghino con la loro autorità alla redintegrazione dell' honor mio."

59. On Galileo's income in this period, see Biagioli, *Galileo's Instruments of Credit*, 7–8.

60. Biagioli, *Galileo's Instruments of Credit*, 7–9.

61. Fenster and Smail, *Fama*, 4; Frank Henderson Stewart, *Honor* (University of Chicago Press, 1994), 9–29.

62. In some matters, such as salaries, the Riformatori's advice needed the civil government's formal approval before it could be carried out; in

this specific case, it seems this was not necessary: the conflict came to an end with the Riformatori's verdict, and that seems to have been the end of it. Grendler, *Universities of the Italian Renaissance*, 157. On the Riformatori dello Studio's role in the censuring of books, see Paul F. Grendler, *The Roman Inquisition and the Venetian Press, 1540–1605* (Princeton University Press, 1977), 151–54.

63. Galileo, *Difesa*, 539.

64. Galileo, *Difesa*, 539–41.

65. Galileo, *Difesa*, 541–44.

66. Galileo, *Difesa*, 544.

67. On Sarpi, see Cozzi, *Paolo Sarpi*; and Muir, *Culture Wars of the Late Renaissance.*

68. Galileo, *Difesa*, 544.

69. Cornaro played a crucial role in rallying support for Galileo's cause: While Galileo went to Venice, Cornaro remained in Padua, where he wrote and spoke to at least five others in favor of Galileo, finding several who were willing to testify on his behalf. *GO* 10.173–78, letters 156, 157, and 159.

70. Galileo, *Difesa*, 546: "et più sentendo il medesimo Galilei che alcuni, per detrarre alla sua fama, andavano parlando che poteva essere che 'l Galilei havesse presa la sua inventione dal Fiammingo."

71. On these men, see the Biographies section of the Museo Galileo's Galileo//thek@, accessed 19 December 2024, https://galileoteca.museogalileo.it.

72. In this regard it is interesting to note that Galileo emphasized Capra's Milanese origins when first approaching the Riformatori and in doing so seemed to draw attention to Capra's "otherness." *GO* 10.172.

73. Galileo, *Difesa*, 547–49.

74. Agostino da Mula, Sebastiano Veniero, and Paolo Sarpi were all Venetians, while Antonio Santini was originally from Lucca. Biographies section, Museo Galileo's Galileo//thek@,; Wilding, *Galileo's Idol*, 48.

75. Galileo, *Difesa*, 548: "che per detti gentil uomini potesse fuora esser dato conto della sufficienza di colui che aveva osato di publicar me per usurpatore e sè per vero inventore di quell' opera."

76. Galileo, *Difesa*, 550–59.

77. Galileo, *Difesa*, 560: "onde immediatamente a suon di trombe fu publicata nello Studio di Padova, nell' ora della maggior frequenza de gli Scolari" and "onde con tal operatione si causeria non picciolo scandolo, et intacco alla riputatione del medesimo Galilei, Lettor in tal professione, et allo Studio ancora."

78. Galileo, *Difesa*, 561: "avendone esso per diverse parti di Europa distribuite già 30."

79. Wilding, *Galileo's Idol*, 45.

80. Galileo, *Difesa*, 532, 600.

81. Wilding, *Galileo's Idol*, 47–48. He also points out that Galileo again relied on his network of authoritative witnesses to disseminate the book, sending them copies to distribute among their own networks.

82. Marr, *Between Raphael and Galileo*, 54.

83. Giovanni Camillo Gloriosi to Johannes Schreck, 29 May 1610, *GO* 10.363. This letter is discussed more extensively in the next chapter.

84. Marino Ghetaldi to Galileo, 20 February 1608, *GO* 10.191: "I Sig.ri Reformatori hanno castigato il Capra dinnanzi al popolo assai bene, ma molto più dinnanzi agli inteligenti l'ha mortificato l'apologia di V. S."

85. Cipriano Saracinelli to Galileo, 11 September 1607, *GO* 10.180: "Il libro che è tocco a me, l'ho letto tutto, et per quello che me ne pare, se quell'ardito Capra sapesse saltare all'indietro, credo che lo farebbe molto volentieri: basta che V. S. l'ha gastigato come meritava, havendolo con la sua penna frustato e mandatolo, come si dice a Fiorenza, su l'asino."

86. Alessandro Tadino to Galileo, 29 November 1619, *GO* 12.498; Ludovico Settala to Galileo, 16 December 1620, *GO* 13.52–53. Also see Camerota, *Galileo Galilei*, 129.

87. Giacomo Alvise Cornaro to Galileo, 24 April 1607, *GO* 10.175: "questo gallante giovane ha trovato una bella via da farsi fammoso con le fatice d'altri; ma la famma potria, di buona et honorata ch'egli pretendea, cangiarsi in rea et vituperosissima."

2. COMPETITION

Some of the material in this chapter previously appeared in Anna-Luna Post, "Putting Galileo in His Place: Geographical Origins and the Rhetoric of Scholarly Credibility," *Renaissance Studies* 37, no. 4 (September 2023): 481–97.

1. A week after the book's appearance, Galileo wrote to the grand duke's secretary, Belisario Vinta, in Florence claiming all copies had "gone away." As Nick Wilding has recently suggested, by this he most likely meant that they had been sent across the Alps, possibly to the book fair in Frankfurt. Wilding, *Galileo's Idol*, 109–10.

2. Henry Wotton to Robert Cecil, 13 March 1610, *GO* app 2.47.

3. Galileo Galilei, *Sidereus Nuncius; or, the Sidereal Messenger*, trans. Albert Van Helden (University of Chicago Press, 1989). For a good introduction to Renaissance astronomy, see William Donahue, "Astronomy," in *The Cambridge History of Science*, vol. 3, *Early Modern Science*,

ed. Katherine Park and Lorraine Daston (Cambridge University Press, 2006).

4. Albert Van Helden, "Galileo and the Telescope"; and Eileen Reeves, "Complete Inventions: The Mirror and the Telescope," both in *The Origins of the Telescope*, ed. Albert Van Helden, Sven Dupré, Rob van Gent, and Huib Zuidervaart (KNAW Press, 2010).

5. Thomas Harriot, in England, beat Galileo to it, while Simon Marius, in Germany, turned his telescope to the sky around the same moment as Galileo. See Pumfrey, "Harriot's Maps of the Moon," 163; and Pasachoff, "Simon Marius's *Mundus Iovialis*," 220–21.

6. H. Floris Cohen, *How Modern Science Came into the World: Four Civilizations, One 17th-Century Breakthrough* (Amsterdam University Press, 2010), 483–84; Paolo Gualdo to Galileo, 29 July 1611, *GO* 11.165.

7. Many contemporaries approached Galileo to try and obtain one of his telescopes, but he rarely complied. Biagioli, *Galileo's Instruments of Credit*, 85–101.

8. Henry Wotton to Robert Cecil, 13 March 1610, *GO* app. 2.47.

9. Biagioli, *Galileo Courtier*; Biagioli, *Galileo's Instruments of Credit*.

10. David Livingstone, *Putting Science in Its Place: Geographies of Scientific Knowledge* (University of Chicago Press, 2003), 16.

11. Anthony Grafton, "A Sketch Map of a Lost Continent: The Republic of Letters," *Republics of Letters: A Journal for the Study of Knowledge, Politics, and the Arts* 1 (1 May 2009): 3; Peter Burke, "Erasmus and the Republic of Letters," *European Review* 7, no. 1 (1999): 8; Peter Miller, *Peiresc's Europe: Learning and Virtue in the Seventeenth Century* (Yale University Press, 2000); David Norbrook, "Women, the Republic of Letters, and the Public Sphere in the Mid-Seventeenth Century," *Criticism* 46, no. 2 (Spring 2004): 223–40; Goldgar, *Impolite Learning*; Lorraine Daston, *Rivals: How Scientists Learned to Cooperate* (Columbia Global Reports, 2023); Daston, "Ideal and Reality of the Republic of Letters in the Enlightenment."

12. David Lux and Harold Cook, "Closed Circles or Open Networks? Communicating at a Distance during the Scientific Revolution," *History of Science* 36, no. 2 (1998): 186–88.

13. Bucciantini, Camerota, and Giudice, *Galileo's Telescope*, 59–61.

14. Paolo Galluzzi, *The Lynx and the Telescope: The Parallel Worlds of Federico Cesi and Galileo*, trans. Peter Mason (Brill, 2017), 12; Bucciantini, Camerota, and Giudice, *Galileo's Telescope*, 106–7.

15. Giovanni Camillo Gloriosi to Johannes Schreck, 29 May 1610, *GO* 10.363: "quorum binos a quibusdam aliis perspicilli beneficio prius detectos fuisse, rumor est. Publice fatetur Augustinus a Mula, patritius Venetus, se huiusmodi stellas prius conspexisse, Galilaeoque, de his nul-

lam notitiam habenti, communicasse; rettulit quoque mihi Ill.mus Fuggerus se audivisse, apud Batavos, ubi perspicilli adinventio ortum habuit, observatos etiam fuisse: a quibus forte excitus Galilaeus, ut gloriae et pecuniae lucrum faceret, et si primus non fuerit observator, primus tamen scriptor haberi voluit." I have slightly adapted the English translation in Bucciantini, Camerota, and Giudice, *Galileo's Telescope*, 106–7.

16. Giovanni Camillo Gloriosi to Johannes Schreck, 29 May 1610, *GO* 10.363: "Quo in crimine Galilaeus suspectus est, cum auctorem quoque se faciat instrumenti quod Circinum Militare et Geometricum appellavit, Magnoque Hetruriae Principi dedicavit; vetus quippe adinventum, et ab omnibus una voce Michaeli Coigneto Antverpiensi, ut primo inventori, attributum."

17. Giovanni Camillo Gloriosi to Johannes Schreck, 29 May 1610, *GO* 10.363: "Scis enim cautos et industrios esse Florentinos: hincque, occasione arrepta, plurimum dignitatis et commoditatis a Republica Veneta, itemque a Magno Hetruriae Duce, adeptus, se perspicilli et novorum planetarum auctorem et inventorem promulgavit." I have slightly adapted the English translation in Bucciantini, Camerota, and Giudice, *Galileo's Telescope*, 106–7.

18. Giovanni Camillo Gloriosi to Galileo, 27 May 1604, *GO* 10.110; Bucciantini, Camerota, and Giudice, *Galileo's Telescope*, 106. On Peiresc, see Cecilia Rizza, *Peiresc e l'Italia* (Giappichelli, 1965); and Miller, *Peiresc's Europe*.

19. He specifically aimed to present the satellites as "Franco-Medicean" by naming them after various members of the Medici family, including the French queen and queen mother. Bucciantini, Camerota, and Giudice, *Galileo's Telescope*, 157.

20. Nicolas Fabri de Peiresc to Giulio Pace, 21 June 1611, *GO* app. 2.92: "Oultre cella saichant qu'il est Florentin, que s'en peult il esperer autre choze que le commun peult esperer d'ung *squaltrito*, joint que s'il ne se tenoit assuré de quelque baston il n'est pas vraysemblable qu'il est tant attandu depuis les suyvantes observations de crainte d'estre prevenu."

21. Lux and Cook, "Closed Circles or Open Networks?," 186.

22. Jean Dietz Moss, *Novelties in the Heavens: Rhetoric and Science in the Copernican Controversy* (University of Chicago Press, 1993), 9–12. Later, Christian virtue became an important element of ethos as well; see George A. Kennedy, *Classical Rhetoric and Its Christian and Secular Tradition from Ancient to Modern Times*, 2nd and enlarged ed. (University of North Carolina Press, 1999), 179; and Serjeantson, "Proof and Persuasion," 139–40, 147–48.

23. Joep Leerssen, *National Thought in Europe: A Cultural History* (Amsterdam University Press, 2006), 13–22, 52–70; Manfred Beller

and Joep Leerssen, eds., *Imagology: The Cultural Construction and Literary Representation of National Characters; A Critical Survey* (Brill, 2007).

24. Leerssen, *National Thought in Europe*, 56–64.

25. Quoted in Leerssen, *National Thought in Europe*, 56–57.

26. Quoted in Leerssen, *National Thought in Europe*, 58–59.

27. Grendler, *Universities of the Italian Renaissance*, 36–37; David Lines, *The Dynamics of Learning in Early Modern Italy: Arts and Medicine at the University of Bologna* (Harvard University Press, 2023), 36; Martina Hacke, "The Messengers of the Nations of the University of Paris and the Book Trade (Late Fifteenth and Sixteenth Centuries)," in *Early Modern Universities: Networks of Higher Learning*, ed. Anja-Silvia Goeing, Glyn Parry, and Mordechai Feingold (Brill, 2020).

28. Joep Leerssen, "Image," in Beller and Leerssen, *Imagology*, 343–44.

29. On Galileo's salary in Padua and his negotiations with the Grand Duke of Tuscany, see Westfall, "Science and Patronage," esp. 16–17; and Mario Biagioli, "Galileo the Emblem-Maker," *Isis* 81, no. 2 (1990): 230–58.

30. Giovanni Bartoli to Belisario Vinta, 27 March 1610, *GO* 10.306–7: "Inviai subito a Padova la lettera per il Galilei, del quale viene da ogniuno letto et considerato un libro fatto di nuovo, dove mostra d'haver col suo occhiale trovato 4 pianeti di più, et visto un altro mondo nella luna, et cose simili, che danno pastura dilettevole ai professori di quelle scientie, massime per il titolo di Sidera Medicea. Non posso già restar di dire, che da molti di questi signori vien stimato hora ch'egli li habbia burlati, quando diede per secreto quel cannone che era molto vulgare, et che nelle piazze si è venduto sino a 4 o 5 lire, della medesima qualità, come si dice; et molti poi se ne ridono, chiamandoli corrivi, mentre egli ha cercato di fare il fatto suo, come ha fatto, et gli è riuscito con un augumento di 500 fiorini alla sua provisione ordinaria per la sua lettura."

31. On Galileo's first endeavors with the telescope, see Van Helden, "Galileo and the Telescope."

32. Belisario Vinta to Galileo, 5 June 1610, *GO* 10.369; Biagioli, *Galileo's Instruments of Credit*, 41.

33. Biagioli, *Galileo Courtier*, 44.

34. Gianfrancesco Sagredo to Galileo, 13 August 1611, *GO* 11.171–72: "Per aventura incerte et dubbiose. . . . V. S. al presente è nella sua nobilissima patria; ma è anco vero che è partita dal luogo dove haveva il suo bene. Serve al presente Prencipe suo naturale . . . Ma quell'essere in luogo dove l'auttorità degli amici del Berlinzone, come si ragiona, val molto, molto ancora mi travaglia." I have adapted the English translation in Bucciantini, Camerota, and Giudice, *Galileo's Telescope*, 194–95.

35. The joke is discussed in Biagioli, *Galileo Courtier*, 251; Bucciantini, Camerota, and Giudice, *Galileo's Telescope*, 197; and—most thoroughly—in Wilding, *Galileo's Idol*, 61–71.

36. See Wilding, *Galileo's Idol*; Cozzi, *Paolo Sarpi*; and Muir, *Culture Wars of the Late Renaissance*.

37. Biagioli, *Galileo Courtier*, 32, 252–53.

38. Sagredo to Galileo, 15 December 1612, *GO* 11.447: "In Padova non si è provisto di Mattematico, perchè li SS.ri Rifformatori vorrebbono uno che havesse letto in altri Studii et fosse huomo di gran fama, et all'incontro dissegnano pagarlo come principiante. . . . Si è sparsa fama ancora che V. S. Ecc.ma, provando costì l'aria et alcun'altra cosa contraria, si ridurrebbe da nuovo in Padova; et io, per ogni buon rispetto, mi son in molti luoghi affatticato di persuadere diversi che questo sarebbe il meglio che potesse occorrere per honorevolezza dello Studio: ma certo che, sì come io trovo compagni in lodarla e stimarla, così in questo particolare della sua ricondotta non è possibile credere il disgusto che gli huomini dimostrano per la sua partenza, et molto più ancora per la maniera che viene detto essere stata tenuta nel partirsi."

39. Sagredo to Galileo, 13 August 1611, *GO* 11.172.

40. On the difficulties observers encountered when using early telescopes, see Albert Van Helden, "Telescopes and Authority from Galileo to Cassini," *Osiris* 9 (1994): 11–12; and Albert Van Helden, "The Telescope in the Seventeenth Century," *Isis* 65, no. 1 (1974): 38–58.

41. English translation from Bucciantini, Camerota, and Giudice, *Galileo's Telescope*, 89–94, (quote on 92); Martin Horky to Johannes Kepler, 27 April 1610, *GO* 10.342–43.

42. English translation from Bucciantini, Camerota, and Giudice, *Galileo's Telescope*, 89–94, (quote on 93); Giovanni Antonio Magini to Johannes Kepler, 26 May 1610, *GO* 10.359.

43. Martin Hasdale to Galileo, 31 May 1610, *GO* 10.365: "Quello, ho da dire a V. S. Ecc.ma, e questo per suo particolare, che oltre l'havere il Magino scritto al Matematico di Colonia per tirarlo alla sua contro di lei, ha fatto il medesimo con tutti i matematici di Germania, Francia, Fiandra, Polonia, Inghilterra ecc.; il che ho saputo non da uno, ma da diversi di diverse nationi, tutte persone che rappresentano persone de principi: dicono agenti residenti, ambasciatori, chè pochi sono in questa Corte con quali non ho qualche intratura o domestichezza (il che sia detto senza ostentatione)."

44. Bucciantini, Camerota, and Giudice, *Galileo's Telescope*, 90; Martin Hasdale to Galileo, 28 April 1610, *GO* 10.344: "Io non ho potuto contenermi di dire che questa non era altro che una mera invidia, perchè biasimano l'opra senza havere visto lo stromento; et che già il pronostico

del Chepplero comminciava a riuscire, perchè dispiace al Magino che altri gli metta il piè avanti, tanto più nella sua patria propria; chè se altrove fosse seguito, meno gli brusciarebbe."

45. Martin Hasdale to Galileo, 31 May 1610, *GO* 10.365: "Ciò è che questo huomo, il Magino, vedendosi mettere il piè inanzi nella propria patria, in quella propria professione dove egli vorrebbe egli solo Fenice, faccia ogni sforzo di scancellare i meriti di V. Ecc.za, in materia et soggetto ella sola merita nome di Fenice."

46. Grendler, *Universities of the Italian Renaissance*, 28–29.

47. Lilti, *Invention of Credibility*, 5; Frisch, *Invention of the Eyewitness*, 11–19.

48. Wilding, *Galileo's Idol*, 137.

49. For a discussion of the initial, more negative reception of Galileo's discoveries among the Lincei, see Galluzzi, *Lynx and the Telescope*, 10–11, 25–29.

50. Galileo Galilei and Christoph Scheiner, *On Sunspots*, trans. and intro. Eileen Reeves and Albert Van Helden (University of Chicago Press, 2010), 1–6.

51. Federico Cesi to Galileo, 3 November 1611, *GO* 11.422–23: "La libertà ch'ella mi porge, mi dà ardire di dirle che non mi pare sia bene in alcun modo tacciar la nazione, ma sì ben la persona e la classe, sotto mano. La nazione è amicissima delle lettere e letterati, e colla moltiplicità de' libri e stampe sostiene la gloria di quelli, e i Lincei particolarmente deveno haverla amica: sono liberi nel filosofare, et vedo honorano molto l'Italiani, mentre non hanno particolar passione o invidia." For Cigoli's letter, see *GO* 11.424–25.

52. Federico Cesi to Galileo, 3 November 1612, *GO* 11.423: "E son sicuro che le sue opre li saranno stimate conforme al dovere, et haveranno altro honore che quelle d'Apelle, ancorchè ei sia della nazione."

53. Galilei and Scheiner, *On Sunspots*, 247.

54. Federico Cesi to Galileo, 22 February 1613, *GO* 11.483–84: "La lettera di V. S. ultimamente riceuta tiene perplesso me con gl'altri Lincei, ch'hora qui si trovano, circa la prefazione del'opra. Lodiamo il consiglio suo, ma il bisogno che vediamo di sbarbare dalle persone indifferenti (de' quali è molto maggior il numero che dell'amici et aversarii di V. S. insieme presi) le cose seminate da gl'invidi et altri aversarii, che vengono defraudandola de' suoi fatti, non ci lascia concorrere afatto seco. Pochi sono di sana e leal mente; e di questi anco pochi in Germania, Francia, Fiandra, anzi qui vicino in Napoli, hanno giusto ragguaglio de' sucessi delli celesti scoprimenti. I suoi libri non sono andati per tutto: V. S. non ha stampato ogni cosa. Li so dire io di certo che molti hanno in tali luoghi mostre le cose da V. S. scoperte; e se alcuni di loro non ardivano

appropriarsele affatto, pur di V. S. non facevano parola. Si pò la prefazione ridurre più grave, si pò con meno affetto e minor dimostrazione far l'istesso effetto. . . . L'epistola dedicatoria secondo l'avertimento si smagrirà un poco. C'era pensiero di mettere un epigramma in lode di Firenze, per piccar sottomano i suoi aversarii."

55. Galilei and Scheiner, *On Sunspots*, 373.

56. The dedication was written by Angelo de Filiis. Galileo Galilei, *Istoria e dimostrazioni intorno alle macchie solari e loro accidenti comprese in tre lettere scritte a Marco Velseri* (Rome, 1613), in *GO* 5.71–249, 77: "E che meraviglia n'è, s'oltre il conoscimento de' meriti, il legame dell'amicizia, col quale egli l'ama ammira ed osserva, la Lince, la patria, l'assidua compagnia li congiungono insieme? La nobil città di Fiorenza, fertile tanto di virtuosi ingegni, ricettacolo insigne di dottrina, che sempre in ogni virtù ha fiorito e fiorisce, ben ragion era che de' proprii frutti e de' suoi scoprimenti prima gustasse e godesse."

57. Kathryn Murphy and Anita Traninger, "Introduction: Instances of Impartiality," in *The Emergence of Impartiality*, ed. Kathryn Murphy and Anita Traninger (Brill, 2013).

58. Johannes Kepler, *Kepler's Conversation with Galileo's Sidereal Messenger*, trans. and intro. Edward Rosen (Johnson Reprint, 1965).

59. On Medici diplomacy and the way Galileo made use of the geographical distance between Florence and Prague to gain Kepler's approval, see Biagioli, *Galileo's Instruments of Credit*, 33–44.

60. Dietz Moss, *Novelties in the Heavens*, 65–96.

61. Kepler, *Kepler's Conversation*, comment by Rosen, 13. Near the end of the dedication, he is referred to as a Medici client. This was not technically true at the time: Galileo's appointment as court mathematician and philosopher would only be granted after Cosimo II received Kepler's endorsement of Galileo's discoveries, as Biagioli has convincingly shown. Kepler's assessment of Galileo's work was thus instrumental in at least one regard: It strongly influenced the Medici's decision to take Galileo on at court. Biagioli, *Galileo's Instruments of Credit*, 39–42.

62. Kepler, *Kepler's Conversation*, comment by Rosen, 6.

63. Kepler, *Kepler's Conversation*, comment by Rosen, 6–7.

64. Bucciantini, Camerota, and Giudice, *Galileo's Telescope*, 108–10.

65. Dietz Moss points out that Kepler had more reasons to feel Galileo had not treated him well; see Dietz Moss, *Novelties in the Heavens*, 87.

66. On Magini's changing attitude, see Galluzzi, *Lynx and the Telescope*, 15–16.

67. The work is reprinted in *GO* 3.1.129–45: Martin Horky, *Brevissima peregrinatione contra Nuncium sidereum* (Modena, 1610).

68. Giovanni Antonio Magini to Antonio Santini, 22 June 1610, *GO* 10.377: "Et io le ho detto che la licentia datoli non è per lui solo, ma per tutti i Tedeschi, che sono inimici di noi altri Italiani."

69. Giovanni Antonio Roffeni to Galileo, 22 June 1610, *GO* 10.376: "assicurandola che se per sorte costui fosse tanto ostinato, come essere sogliono li Tedeschi"; and 29 June 1610, *GO* 10.384: "et il detto sempre ha cercato levarlo di questo pensiero; ma in soma li oltramontani sono zervelli molto stravaganti." This same friend published a treatise refuting Horky in 1611: Giovanni Antonio Roffeni, *Epistola apologetica contra caecam peregrinationem Cuiusdam furiosi Martini, cognomine Horkij editam adversus nuntium sidereum: De quattuor novis planetis Gallilei Gallilei olim in Patavino Gymnasio publici Mathematici* (Bologna, 1611), in *GO* 3.1:191–200. The Italian letter on which the Latin treatise was based was already circulating in manuscript form in September and October 1610.

3. ADMIRATION AND APPROPRIATION

1. Robert Darnton, *The Forbidden Best-Sellers of Pre-Revolutionary France* (Norton, 1996); Margócsy, *Commercial Visions*; Goldgar, *Impolite Learning*, 150–54. On portraiture and scholarship, see especially Patricia Fara, "Framing the Evidence: Scientific Biography and Portraiture," in *The History and Poetics of Scientific Biography*, ed. Thomas Söderqvist (Routledge, 2007).

2. Crystal Hall, *Galileo's Reading* (Cambridge University Press, 2013), 3–4. See also Claire Preston, *The Poetics of Scientific Investigation in Seventeenth-Century England* (Oxford University Press, 2015).

3. Braudy, *Frenzy of Renown*, 252–53, 263.

4. Heilbron, *Galileo*, 60–62; Stillman Drake, "Galileo Gleanings IV: Bibliographical Notes," *Isis* 49, no. 4 (December 1958): 409–13.

5. Biagioli, *Galileo Courtier*, 129–34.

6. Galileo to Belisario Vinta, 19 March 1610, *GO* 10.300: "Questa seconda volta credo che lo farò in lingua toscana, sì perchè, oltre a i librai, ne sono pregato da molti altri, sì ancora perchè credo che le Muse toscane non taceranno in così grande occasione le glorie di questa Ser.ma Casa, perchè sin qua sono alcuni che scrivono in questo proposito: et tali componimenti si potranno prefigere all' opera."

7. Alessandro Sertini to Galileo, 7 August 1610, *GO* 10.412: "V. S. non mi ha mai detto cosa alcuna dello stampare."

8. Biagioli, *Galileo Courtier*, 136–37.

9. Sertini's friendship with Galileo went back to at least 1593, when Sertini sent Galileo a poem in honor of his (Sertini's) dead brother and asked Galileo's opinion on it. He also requested a poem in return, either

by Galileo or someone else. Alessandro Sertini to Galileo, 19 November 1593, *GO* 10.63–64.

10. Only Seripandi did not have a membership in either academy. Alison Brown, "Defining the Place of Academies in Florentine Culture and Politics," in *The Italian Academies 1525–1700: Networks of Culture, Innovation and Dissent*, ed. Jane E. Everson et al. (Routledge, 2016), 22; Biagioli, *Galileo Courtier*, 107n16, 136–37. Galileo would serve as consul of the Accademia Fiorentina between 1620 and 1623. *GO* 19.444, 221.

11. Alessandro Sertini to Galileo, 29 June 1610, *GO* 10.306: "Il Sig.r Andrea scrive di nuovo a V. S.; sì che io non so che me li dire di lui, se non ch'egli gl'è servitore. Le Muse vanno un poco adagio, perchè le nove sono rimaste indietro per una decima che debbe haver più bel muso. V. S. bisogna che gli scriva da sè, s'ella vuole che faccia qual cosa sopra le stelle Medicee. Io ne ho gettato un motto co 'l Sig.r Buonarruoti; con gl'altri, non havendo tanta familiarità, non so come mi fare, ancho volendo. E perchè è tardi, finisco, e le bacio le mani." Sig. Andrea is probably Andrea Salvadori.

12. Biagioli, *Galileo's Instruments of Credit*, 33–41; *GO* 10.355–56, 369, 372–75, 383–84.

13. Sertini to Galileo, 10 July 1610, *GO* 10.399. The poems that would arrive shortly were by De' Bardi and Buonarroti. Sertini also referred to the finished poems of Niccolò Arrighetti and Claudio Seripandi, mentioned that Andrea Salvadori was still at work, and that Chiabrera had also promised to send something. Chiabrera never sent in a poem for the Tuscan edition of the *Sidereus Nuncius*, although he did refer briefly to Galileo in a later poem. Nunzio Vaccalluzzo, *Galileo nella poesia del suo secolo* (Remo Sandron, 1910), xxviii–xxix, discusses this letter as well as Chiabrera's other poetic endeavors and concludes Chiabrera may have hesitated to honor Galileo in print because he had been educated by the Jesuits and thought it more prudent to remain silent.

14. Sertini to Galileo, 10 July 1610, *GO* 10.399: "Non so come questi signori se l'intendino circa 'l mettere il lor nome, caso che V. S. le voglia stampare: intenderò l'umor loro."

15. Richard McCabe, *"Ungainefull Arte": Poetry, Patronage, and Print in the Early Modern Era* (Oxford University Press, 2016), 58–60; Braudy, *Frenzy of Renown*, 309.

16. Sertini to Galileo, 7 August 1610, *GO* 10.412: "E perchè V. S. disse di volere stampare, ogn'uno ne ha paura, ed egli ancora non vorrebbe il suo nome in istampa, ma come il Sig.r Piero de' Bardi, havendosi a stampare, si contenterebbe che si dicesse: dell'Impastato, Accademico della Crusca . . . V. S. non mi ha mai detto cosa alcuna dello stampare." See also Vaccalluzzo, *Galileo nella poesia*, xxix.

17. Sertini to Galileo, 7 August 1610, *GO* 10.411; Biagoli, *Galileo Courtier*, 138–39; Bucciantini, Camerota, and Giudice, *Galileo's Telescope*, 199.

18. Asdrubale di Montauto to Cosimo II, 31 July 1610: "Si sente che venghino fora scritture da più bande contro al matematico Galilei e in particolare di verso Praga un libro et che haverà molto da fare." Archivio di Stato di Firenze (ASF), Mediceo del Principato, vol. 3001, fol. 279r, available through the Medici Archive Project's online database, the BIA platform.

19. On early modern paratext, see Helen Smith, ed., *Renaissance Paratexts* (Cambridge University Press, 2011). For a more extensive discussion on preliminary poems, see Harm-Jan van Dam, "Poems on the Threshold: Neo-Latin *carmina liminaria*," *Acta Conventus Neo-Latini Monasteriensis: Proceedings of the Fifteenth International Congress of Neo-Latin Studies*, ed. Astrid Steiner-Weber and Karl A. E. Enenkel (Brill, 2015), 53.

20. Vaccalluzzo, *Galileo Galilei nella poesia*, 57–58, 59–60.

21. Sertini to Galileo, 7 August 1610, *GO* 10.412: "Il Sig.r Buonarruoti le bacia le mani e le manda l'alligata composizione, pregandola che vuoglia migliorarla dove le paia che ne sia capace, e che le piaccia aggradire la buona volontà di servirla. Credo, anzi son certo, che le piacerà." For a particularly poignant example of this practice, see Meredith Ray's discussion of the development of Margherita Sarrocchi's long poem *Scanderbeide*; Sarrocchi told Galileo by way of a letter that she had intentionally left blank some spaces in the poem so that she could decide later whom she could best honor by referring to them in her work. Meredith Ray, *Margherita Sarrocchi's Letters to Galileo: Astronomy, Astrology, and Poetics in Seventeenth-Century Italy* (Springer, 2016), 31–32.

22. In August 1610, he had already observed what he thought were two bodies circling around Saturn—an important new discovery he would need to incorporate into the translation. As Galileo's catalog of new findings grew over the course of 1610 and 1611, it became more and more logical to publish a new book, rather than a translation of the work that had gained Galileo his newfound fame. Bucciantini, Camerota, and Giudice, *Galileo's Telescope*, 119–24, 199–200.

23. Galileo to Michelangelo Buonarroti the Younger, 16 October 1610, *GO* 10.446–47; Ludovico da Cigoli to Galileo, 13 November 1610, *GO* 10.475; Buonarroti to Galileo, October 1610, *GO* 10.452–53; Buonarroti to Maffeo Barberini, 22 March 1611, *GO* 11.72. On Buonarroti's brokerage, see Cole, "Cultural Clientelism and Brokerage Networks."

24. Galileo to Michelangelo Buonarroti the Younger, 26 June 1638, *GO* 17.346–47. See Tognoni, *GO* app. 1.52–53, on the iconography and

execution of the ceiling's design; Tognoni also discusses a deeply ambiguous poem that Buonarroti briefly considered including beneath the portrait. Its verses indicate that he clearly sided with Galileo's heliocentric stance.

25. Vaccalluzzo, *Galileo Galilei nella poesia*, 3–10. For Andrea Salvadori's biography, see the entry for him in *Dizionario Biografico degli Italiani*, vol. 89 (2017), at Treccani, http://www.treccani.it/.

26. Biagioli, *Galileo Courtier*, 139n129.

27. Biblioteca Nazionale Centrale di Firenze (BNCF), Fondo Galileiano ms. 28 II, Galileo. I.18, Lavori letterari, 4r–18r.

28. On Magagnati, see Antonio Favaro, *Amici e corrispondenti di Galileo*, ed. Paolo Galluzzi (Libreria editrice Salimbeni, 1983), 65–92; Stillman Drake, "Galileo Gleanings XIV: Galileo and Girolamo Magagnati," *Physis* 6 (1964): 269–86; Bucciantini, Camerota, and Giudice, *Galileo's Telescope*, 189–191; and Andrea Battistini, *Galileo e i Gesuiti: Miti letterari e retorica della scienza* (Vita e Pensiero, 2000), 69–76; Andrea Battistini, "'Cedat Columbus' e 'Vicisti, Galilaee!': Due esploratori a confronto nell'immaginario barocco," *Annali d'Italianistica* 10 (1992): 123.

29. The friend was Giovanni Contarini, a painter with whom Galileo also exchanged letters. Carlo Carabba and Giuliano Gasparri, "La vita di Girolamo Magagnati," in Girolamo Magagnati, *Lettere a diversi*, ed. and intro. Laura Salvetti Firpo (Leo S. Olschki, 2006), vii– xx, xi; Carlo Ridolfi, *Le Meraviglie dell'Arte overo le vite de gl'illustri pittori Veneti, e dello stato*, vol. 2 (Venice, 1648), 94.

30. Carabba and Gasparri, "La vita di Girolamo Magagnati," xii–xiii.

31. Favaro, *Amici e corrispondenti*, 443–65, 448–49.

32. Crystal Hall, "Galileo, Poetry, and Patronage: Giulio Strozzi's *Venetia edificata* and the Place of Galileo in Seventeenth-Century Italian Poetry," *Renaissance Quarterly* 66, no. 4 (2013): 1301.

33. Girolamo Magagnati, *Meditazione poetica sopra i pianeti medicei* (Venice, 1610).

34. Vaccalluzzo, *Galileo Galilei nella poesia*, 11–12. For a discussion of how Galileo's subsequent publications repeated this visual theme, see Karl Enenkel, *Die Stiftung von Autorschaft in der neulateinischen Literatur (ca. 1350–ca. 1650): Zur autorisierenden und wissensvermittelnden Funktion von Widmungen, Vorworttexten, Autorporträts und Dedikationsbildern* (Brill, 2015), 135, 428.

35. Magagnati, *Meditazione*, 5: "Ma fine avran pietre, metalli, e monti, / Sudditi al Tempo, e al tenebroso Oblio; / Vedrà stampata ne' celesti annali / Frà gli Astri fiammeggiar lucente, e pura / Del tuo ceppo la gloria, e del tuo nome / L'ampia de l'Universo immensa mole."

36. Magagnati, *Meditazione*, 6: "Il Ligure fulgor, che Tifi oscura . . . Nè riportò piu ch'altro Mondo al Mondo: Ma tu solcasti, o Galileo, de l'etra / Gli smisurati inaccesibil campi, / E profondato il curioso aratro / Di spirto vago entro i zaffiri eterni / Rivolvendo del Ciel l'aurate zolle / Ritrovasti nov'Orbi e novi Lumi." For a longer discussion of the poem's themes and imagery, see Battistini, *Galileo e i Gesuiti*, 69–76.

37. Davis, *Gift in Sixteenth-Century France*. For gifts in the specific context of Galileo's patronage relations, see Biagioli, *Galileo Courtier*, 36–53.

38. McCabe, *"Ungainefull Arte,"* 17.

39. Magagnati to Cosimo II, May 1610, *GO* app. 2.55: "ho sentito quasi violentarmi forse' da gl'influssi de Medesimi, a' scriver questi pochi versi."

40. Galileo to Belisario Vinta, 21 May 1610, *GO* 10.354–55. For an overview of Magagnati's poems, see Favaro, *Amici e corrispondenti*, 65–92; and Carabba and Gasparri, "La vita di Girolamo Magagnati," xx.

41. Magagnati to Galileo, summer of 1610, *GO* app. 2.72; Drake, "Galileo Gleanings XIV," 278–84.

42. Girolamo Magagnati to Jacques Badovère, June 1610, *GO* app. 2.59: "Serva per ramemorarle la mia divotione e il desiderio d'ho di servirla, e per pregarla (se le parà cosa non del tutto indegna) di farne capitar una all.'Ill.mo e R.mo S.r Card. Gioiosa per alleviar per mez'ora il peso delle sue gravissime occupazioni, nel qual breve spazio, per aventura, desti qualche attimo di ricordanza della humilissima servitù mia la quale perpetua benché inutile gli conservo." The letter to Chiabrera is in *GO* app. 2.56.

43. On Figliucci, see Charles Edwards O'Neill and José María Domínguez, eds., *Diccionario histórico de la Compañía de Jesús*, vol. 2 (Institutum Historicum, SI, and Universidad Pontificia Comillas, 2001), 1426.

44. The poem was entitled "Sonnet on the Death of the King Henry the Great, and on the Discovery of Some New Planets, or Stars Wandering Around Jupiter, Made This Year by Galileo Galilei, Famous Mathematician of the Grand Duke of Florence." Bucciantini, Camerota, and Giudice, *Galileo's Telescope*, 162–63.

45. Freedberg, *Eye of the Lynx*, 105–11.

46. Luigi Guerrini, "Le *Stanze sopra le stelle e macchie solari scoperte col nuovo Occhiale* di Vincenzo Figliucci: Un episodio poco noto della visita di Galileo Galilei a Roma nel 1611," *Lettere Italiane* 50, no. 3 (1998): 401.

47. Mordechai Feingold, "*Fama*: Les savants jésuites et la quête de la renommée," *Dix-Septième Siècle* 237, no. 4 (2007): 755–74.

48. Lorenzo Salvi [pseud. of Figliucci], *Stanze sopra le stelle e macchie solari scoperte col nuovo Occhiale* (Rome, 1615), 24–34. To avoid confusion and reflect Figliucci's authorship, I will from here on refer to this work as Figliucci, *Stanze sopra le stelle.*

49. Guerrini, "Le *Stanze*," 393.

50. Figliucci, *Stanze sopra le stelle*, 9: "Tu, Galileo, sopra terrestre limo / Il sentier chiuso a noi primiero apristi. / Tu co' i cristalli, ch'io ne'carmi esprimo, / Di nuove stelle il ciel ricco scopristi; / Mentr'altri al terreo suol, tu il cor alzasti / A merci eterne, e 'l mar del ciel solcasti."

51. Battistini, *Galileo e i Gesuiti*, 71–72. See also the poems by Paganino Gaudenzi, written after Galileo's death and discussed in chapter 5 of this book, which praise him specifically for the noble character of his objects of research: *In morte del famosissimo Galileo tre sonetti* (N.p., 1642).

52. Figliucci, *Stanze sopra le stelle*, 9.

53. Figliucci, *Stanze sopra le stelle*, 9; Guerrini, "Le *Stanze*," 403–4. The dedication is to Cardinal Aldobrandini, Camerlengo, and must refer to Pietro Aldobrandini, who was camerlengo from 1599 until his death in 1621.

54. Guerrini, "Le *Stanze*," 403–4. For an English translation of several stanzas, as well as a more extensive discussion of Figliucci's ties with the Collegio Romano, see Bucciantini, Camerota, and Giudice, *Galileo's Telescope*, 205–7, 211.

55. Camerota, *Galileo Galilei e la cultura scientifica*, 257. For the conflict over the sunspots, see Galilei and Scheiner, *On Sunspots.*

56. The town had offered fierce resistance to all attempts by the Medici family to extend Florence's power over it and had only become subject to Medici rule in 1557, when the city surrendered to the Spanish king, Philip II, who subsequently gifted the city to the Medici family.

57. Figliucci, *Stanze sopra le stelle*, 41: "Ma chi nel Sole a rimirar intento, / Recò primiero a noi di ciò novella? / Dubbio ancor prende, e non lusingo, o mento, / Che due fin'hor il mondo incerto appella. / Te lodar la Germania altiera sento, / Apelle, e celebrar l'opra si bella. / Di te Fiorenza, Galileo, si vanta, / E 'l tuo nome superba honora, e canta. // Ma d'egual merto, e d'egual lode appare / L'un e l'altro di lor degno, a chi pensa / Come l'un senza l'altro ardì fissare / Lo sguardo infermo in quella luce immensa. / Quei dal Reno al Danubio le fa' chiare, / E 'l bel segreto a' suoi German dispensa. / Questi di qua da l'Alpe a noi le mostra / A l'Arno, al Tebro, a l'alma Italia nostra."

58. Figliucci, *Stanze sopra le stelle*, 9: "Tu co' i cristalli, ch'io ne'carmi esprimo, / Di nuove stelle il ciel ricco scopristi."

59. Cesi to Galileo, 1 March 1614, *GO* 12.28: "Intesi qui in una conversatione che un poeta moderno (credo barzellettista, benchè nè anco

potei intenderne il nome) componeva sopra i nuovi Pianeti in lode d'un Principe, alludendo con essi (non altrimente che s'egli ci havesse qualche ius sopra) al'arme di quello stellata, servendosene a suo modo, senza nomarli Medicei."

60. Cesi to Galileo, 2 February 1615, *GO* 12.136: "M'è ben hora appunto stato mandato di Roma un'operetta di stanze sopra le stelle e macchie solari scoperte col nuovo occhiale. L'authore di questa è un Sig.r Lorenzo Salvi, gentilhuomo Senese. Non l'ho ancor veduta, se non che in una guardata ho visto che parla anco di V. S., ma non quanto si converebbe, e mette Appelle a parte nel'invention delle machie."

61. Luigi Maraffi to Galileo, 12 December 1615, *GO* 12.209–10: "Quello che pare a me, è che molto scarsamente sia proceduto con la lode dove et con chi la meritava, tanto più che, vestendosi da poeta, poteva maggiormente allargarsi. Inculca più volte che l'occhiale è stato trovato in Fiandra, migliorato in Italia, ma non dice da chi; che con l'occasione delle stelle di Giove altri hanno osservate altre stelle, come sono i matematici del Collegio Romano Giesuiti; che il primo osservatore delle macchie solari è dubbio chi sia, ma però che la sta nel finto Apelle Giesuita et in V. S.; et perchè debbe havere la procura dalle parti, si fa arbitro, et giudica che l'uno et l'altro è il primo, ma uno in Germania et l'altro in Italia. Dove parla delle stelle intorno a Giove (le quali mai, che io mi ricordi, chiama Medicea), dice pure che l'inventione è di V. S.; et quanto dice e s'allarga è questo poco d'ottava, dalla quale vedrà, come da uno saggio, la S. V. la qualità del verso." On the rivalry between the Dominican and Jesuit orders, see Mordechai Feingold, "The Grounds for Conflict: Grienberger, Grassi, Galileo, and Posterity," in *The New Science and Jesuit Science: Seventeenth Century Perspectives*, ed. Mordechai Feingold (Springer, 2003).

62. Freedberg, *Eye of the Lynx*, 65–70; Galluzzi, *Lynx and the Telescope*, chap. 2; Christoph Lüthy, "Atomism, Lynceus, and the Fate of Seventeenth-Century Microscopy," *Early Science and Medicine* 1, no. 1 (1996): 7.

63. Elisja Schulte van Kessel, *Geest en Vlees in godsdienst en wetenschap* (Staatsuitgeverij, 1980), 153; Freedberg, *Eye of the Lynx*, 66–69.

64. For a more elaborate discussion of the academy's early struggles, see Van Kessel, *Geest en Vlees*; Freedberg, *Eye of the Lynx*; and Richard S. Westfall, "Galileo and the Accademia dei Lincei," in *Novita celesti e crisi del sapere: Atti del Convegno internazionale di Studi galileiani*, ed. Paolo Galluzzi (Giunta Barbèra, 1984), 195–96.

65. Freedberg, *Eye of the Lynx*, 72–73.

66. Freedberg, *Eye of the Lynx*, 101–2; Galluzzi, *Lynx and the Telescope*, chap. 1.

67. Van Helden, "Invention of the Telescope"; Van Helden et al., *Origins of the Telescope.*

68. Freedberg, *Eye of the Lynx*, 105.

69. Freedberg, *Eye of the Lynx*, 108–12.

70. As Richard Westfall has noted, Cesi's relation to Galileo differed slightly from the way he approached the other Lincei: Cesi signed his letters to Stelluti and Faber "Principe di Lincei," whereas he closed his letters to Galileo with "always most ready to serve you, Federico Cesi, Marquis of Monticelli." Westfall, "Galileo and the Accademia dei Lincei," 197.

71. Cesi to Stelluti, April 1613: "Non si cerchi luogo celebre per il Liceo, ma secondo il Linceografo, chè i Lincei il nome, honore e fama hanno d'haverlo e ottenerlo solo con libro et opre." Giuseppe Gabrieli, ed., "Il Carteggio Linceo della vecchia Accademia di Federico Cesi (1603–1630)," *Memorie della R. Accademia dei Lincei*, series 6, Classe di scienze morali, storiche e filologiche, 7 (1938–1942), part II (1610–1624), 350–51.

72. The academy's first publication, a treatise on the supernova of 1604 by Johannes van Heeck, had already been published by 1605, and two books by Della Porta followed in 1610. Galluzzi, *Lynx and the Telescope*, 39.

73. Galileo first responded to Scheiner's *Tres Epistolae de Maculis Solaribus* (Augsburg, 1612). After Galileo had written two letters to Welser, Scheiner published his *De Maculus Solaribus et Stellis circa Iovem Errantibus Accuratior Disquisitio* (Augsburg, 1612). Galileo's third letter is a reply to this second work by Scheiner. See Galilei and Scheiner, *On Sunspots*, for a detailed discussion of all texts and their precise chronology.

74. Cesi to Galileo, 3 November 1612, *GO* 11.422–423, and 22 February 1613, *GO* 11.483–84.

75. For accounts of the many ways in which the Lincei were involved in the different stages of book publishing, see Sabina Brevaglieri, "Science, Books and Censorship in the Academy of the Lincei," in *Conflicting Duties: Science, Medicine and Religion in Rome (1550–1750)*, ed. Maria Pia Donato and Jill Kraye (University of London Press, 2009), 109–33. On Cesi's work on Van Heeck's treatise of 1604, see Freedberg, *Eye of the Lynx*, 91–98.

76. Translation by Reeves and Van Helden in Galilei and Scheiner, *On Sunspots*, 378–79. The original Latin reads as follows: "Dum radio, GALILAE, tuo coelum omne retectum / Spectat, et insolito murmure Terra fremit; / Quod contra tempus solido non aere resistit / Aeterna in fragili stat Tibi fame vitro." *GO* 5.91.

77. One of Cesi's letters refers to an "elegia" Valerio wrote to include in this work, which would be printed "to the mortification of your enemies." Cesi explicitly asked Galileo for his opinion on the elegy and prodded him to break his silence (Galileo was often slow to respond). Galileo's response is lost, but it seems unlikely that the poem that was eventually printed is the one Cesi referred to, as the short epigram is not particularly biting in tone or content. Cesi to Galileo, 2 March 1613, *GO* 11.487: "Il S.r Valerio non s'è potuto contenere di non far l'inclusa elegia, perchè a mortificatione degl'aversarii di V. S. si stampasse. Non s'è fatt'altro senza che V. S. non ne gusti: e veramente non possiamo approvare affatto il tacere; pure V. S. giudichi e commandi."

78. Galileo at first had wanted to dedicate his work to Cesi; Cesi instead encouraged him to dedicate it to the grand duke. Eventually, they settled on Salviati as the dedicatee of the work. Galluzzi, *Lynx and the Telescope*, 102–3.

79. I have slightly adapted the translation of Reeves and Van Helden used in Galilei and Scheiner, *On Sunspots*, 379. The original Latin reads as follows: "Non tibi Daedalis opus est, GALILAEE, volanti / Ad solem pennis; Sole tepente cadunt. / Nec Ganymedaea veheris super astra volucri: / Imbelles pueros haec modo portat avis. / Ast tibi, ceu LYNCI, penetrent quae moenia coeli, / Lumina praeclarum contulit ingenium, / Queis nova demonstras tu sidera PRIMUS Olympo, / Atque subesse novas Sole doces Maculas." *GO* 5.91.

80. On the figure of Daedalus in the early modern imagination, see Ronny F. Schulz, "Myths of the Inventor: Inventing Myths in the Literary Concept of the Artistic Ingenium in Germany and Italy (1500–1550)," in *Allusions and Reflections: Greek and Roman Mythology in Early Modern Europe*, ed. Elisabeth Waghäll Nivre et al. (Cambridge Scholars, 2015). On the connection between Daedalus, Icarus, mathematics, and instruments, see Deborah Harkness, *The Jewel House: Elizabethan London and the Scientific Revolution* (Yale University Press, 2008), 97–141.

81. Sarah Carter, *Ovidian Myth and Sexual Deviance in Early Modern English Literature* (Springer, 2011), 87–90; Allison B. Kavey, "Mercury Falling: Gender Flexibility and Eroticism in Popular Alchemy," in *The Sciences of Homosexuality in Early Modern Europe*, ed. Kenneth Boris and George S. Rousseau (Routledge, 2008).

82. Translation by Reeves and Van Helden from Galilei and Scheiner, *On Sunspots*, 379–80. The original poem in full reads as follows: "Son, GALILEO, tuoi pregi or sì possenti, / Che da la face del notturno orrore / Spuntan, per seggio di tua gloria, fuore / Ben cento Olimpi ad onorarti intenti. / E qualor co' tuoi vetri, industre il tenti, / S'inchinan l'alte spere

a tuo favore; / E per far vie più chiaro il tuo valore, / Nascon a mille a mille orbi lucenti. / L'apportator del giorno ach'ei comparte / Prodigo il lume a te, ch'il fura intanto / Del suo bel volto a la più chiara parte. / Così di macchie asperso il puro manto / Tu primier ce l'additi; e con tal arte / Fregi d'immortal luce il tuo gran vanto." *GO* 5.92.

83. On the early modern branding of authors, see Pettegree, *Brand Luther.*

84. Karl Enenkel has interpreted the Lynx to signify "mental power, reason and intelligence," but of course the most important association the Lincei wished to evoke was with the sharp, piercing eyesight that leads to truthful observations. Enenkel, *Die Stiftung von Autorschaft*, 368–69. Paolo Galluzzi has convincingly traced the development of Cesi's family crest on different publications sponsored by the Accademia dei Lincei: In three stages it evolves from the Cesi family's original weapon, to a small, crowned tree standing on six small hills printed on Van Heeck's 1605 treatise, to the Lincean vignette depicted on the title page of the *Istoria e dimostrazioni*. Galluzzi, *Lynx and the Telescope*, 37–43.

85. Cesi to Galileo, 4 July 1612, *GO* 11.51, informs Galileo that Welser's name has been suggested by the German Lincei (Faber and Schreck). On 7 October, Cesi informed Galileo that Welser had joined the academy. *GO* 11.409.

86. As Reeves and Van Helden have justly pointed out, it is striking that Galileo forgot to include the word *filosofo* in the title he sent Cesi on 5 January, especially in light of his fervent negotiations with the Tuscan court in the spring of 1610. Galilei and Scheiner, *On Sunspots*, 245.

87. On title pages and portraits, see Braudy, *Frenzy of Renown*, 300–5; Peter Burke, "Renaissance Individualism and the Portrait," *History of European Ideas* 21, no. 3 (1995): 393–400; Fara, "Framing the Evidence"; and Alessandro Tosi, *Portraits of Men and Ideas: Images of Science in Italy from the Renaissance to the Nineteenth Century* (Plus, 2007).

88. See Tognoni's discussion of the portrait in *GO* app. 1.34–35; on the costs of producing the portrait, see *GO* 19.266.

89. Stelluti to Galileo, 12 April 1613, *GO* 11.494: "essendo tutti i libri in potere del nostro Bibliotecario, quale, compita la debbita distribuzione a' Lincei et amici, doverà del restante farne fare esito, applicandone il ritratto a benefizio della Compagnia."

90. Galluzzi, *Lynx and the Telescope*, 51. The inscription surrounding Galileo's portrait further emphasizes his membership in the Lincei, while also referring to his affiliation with the Grand Duke of Tuscany.

91. In their translation notes, Reeves and Van Helden argue that the way the *putto* holds the trumpet is "an iconographical touch that does not suggest that Galileo invented the telescope, but rather that he was

the first to achieve renown in deploying it." Galilei and Scheiner, *On Sunspots*, 247.

92. Cesi to Galileo, 2 March 1613, *GO* 11.497: "N'ha mostro particolarissima sodisfazzione e stima, e s'è humanissimamente offerto; di modo che potrà non poco giovare alle nostre cose e al nostro nome in Germania."

93. Cesi to Galileo, 2 March 1613, *GO* 11.487: "Parte il P.e di Bamberga, di qua lunedì, o al più longo giovedì, se bene si tien più sicuro il primo. Verrà a Fiorenza: ho preso ardire d'offerirgli che V. S. le mostrarà i spettacoli celesti: credo senz'altro lo desiderarà: però mi facci gratia farseli conoscere, offerirsele e mostrarli, e potrà dirle che ha saputo da me la sua benignità e da gl'altri Lincei, quali le son tanto servitori etc."

94. Cesi to Galileo, 2 March 1613, *GO* 11.486–88: "Ne vanno con questa occasione 15 al S.r Velsero, presto e sicuri. Ne sono fatti quaranta in circa per questa fretta, nè s'è voluto farne più de' primi fogli, volendo prima saper se le piace in questo modo, e della prefazione, chè molti suoi affettionati vi vorrebbono ad ogni modo qualche cosa."

95. Cesi to Galileo, 24 November 1612, *GO* 11.438, discusses the original plan of printing three thousand copies. His letter to Galileo of 29 September 1612 discusses his wish to pay for the work. *GO* 11.404. Printing Galileo's three letters cost 170 *scudi*, Apelles's work cost 38, and 50 *scudi* were added "per stampatura delle Macchie da finirse." On the printing costs, see *GO* 11.265–66.

96. Adrian Johns, *The Nature of the Book: Print and Knowledge in the Making* (University of Chicago Press, 1998) 25.

97. Cesi to Stelluti, April 1613, in *Atti della Reale Accademia dei Lincei*, 350–51.

98. Stelluti to Galileo, 13 April 1613, *GO* 11.494: "Fra li cento libri ve ne sono dieci separati, di carta più fina, per avviso." In 1623, when Stelluti again sent Galileo copies of one of his works (this time it was his *Il Saggiatore*), they again included a few books printed on different paper. These copies on "carta più fina" were meant for Galileo's friends, which suggests that the 1613 copies on finer paper were meant to be distributed at the court rather than sold for a lower price. Stelluti to Galileo, 28 October 1623, *GO* 13.142: "fra detti libri ve ne sono otto di carta più fina, che serviranno per dare a cotesti SS.ri suoi amici."

99. Pettegree, *Brand Luther*.

100. *Dal Diario di Viaggio di Giovanni Tarde*, in *GO* 19.589–93, 591: "que par toute l'Italie et Alemaigne on l'appelloit *philosophus linceus*."

101. In the meantime, Galileo did publish a treatise on comets, together with Mario Guiducci. This work appeared under Guiducci's name, however, and consequently the Lincei did not take as active an

interest in it as in the works Galileo wrote and published under his own name.

102. Cesarini to Galileo, 8 August 1623, *GO* 13.124; Stelluti to Galileo, 12 August 1623, *GO* 13.121.

103. Translation from Gattei, *On the Life of Galileo*, 304–7.

104. Stelluti to Galileo, 8 September 1623, *GO* 13.129; Galileo to Federico Cesi, 9 October 1623, *GO* 13.134. On Lincei involvement in the printing of *Il Saggiatore*, see Biagioli, *Galileo Courtier*, 289–97.

105. Freedberg, *Eye of the Lynx*, 144–45. Galileo hoped for explicit permission to publish a new work on heliocentrism but did not get it. On Galileo's Roman sojourn of 1624, see Shea and Artigas, *Galileo in Rome*, chap. 4. A succinct discussion can also be found in Heilbron, *Galileo*, 256–59.

106. Susan Barnes, "The Uomini Illustri, Humanist Culture, and the Development of a Portrait Tradition in Early Seventeenth-Century Italy," *Studies in the History of Art* 27 (1989): 84: The original text reads as follows: "li Somme Pontefici . . . li Principi Cardinali, e Signori titolate, e d'ogni altra qualita, pur che famosi fussero, si religiosi come seccolari."

107. Barnes, "Uomini Illustri," 83; *GO* app. 1.41–45.

108. Barnes, "Uomini Illustri," 88. Little is known about the etched portraits' distribution at the time, but the fact that many museums across the globe possess original copies suggests that they may have been sold individually or as sets.

109. Freedberg, *Eye of the Lynx*, 76–77.

4. OPPOSITION

1. Caccini's deposition to the Holy Office, given on 20 March 1615, *GO* 19.307: "Presi per tanto occasione da questo luogo, da me prima in senso litterale et poi in sentimento spirituale, per salute delle anime, interpretato, di riprovare, con quella modestia che conviene all'offitio che tenevo, una certa opinione già di Nicolò Copernico, et in questi tempi, per quel ch'è publichissima fama nella città di Firenze, tenuta et insegnata, per quanto dicono, dal Sig.r Galileo Galilei matematico."

2. Heilbron, *Galileo*, 167–70. On Galileo's trip to Rome, see Freedberg, *Eye of the Lynx*, 108–12; and Shea and Artigas, *Galileo in Rome*, 19–48.

3. Galileo's first draft of the *Discorso* includes a discussion of the first oral dispute and its aftermath; see *Diversi frammenti attenenti al trattato delle cose che stanno su l'acqua* in *GO* 4.17–56.

4. Ku-ming (Kevin) Chang, "From Oral Disputation to Written Text: The Transformation of the Dissertation in Early Modern Europe,"

in *History of Universities*, vol. 19, ed. Mordechai Feingold (Oxford University Press, 2004), esp. 132–33; William Clark, *Academic Charisma and the Origins of the Research University* (University of Chicago Press, 2006), 74–80; Biagioli, *Galileo Courtier*, 55, 60–65.

5. In the draft version of his treatise, Galileo refers only to his encounter with a "professor of philosophy." *Diversi frammenti*, *GO* 4.32. Biagioli, *Galileo Courtier*, 170–71, states that both Vincenzo di Grazia and Giorgio Coresio were present during the first installation of the dispute, but in their printed treatises neither one claimed to have been there.

6. Stillman Drake, *Cause, Experiment, and Science: A Galilean Dialogue Incorporating a New English Translation of Galileo's "Bodies That Stay atop Water, or Move in It"* (University of Chicago Press, 1981), xv–xxix; William Shea, "Galileo's Discourse on Floating Bodies: Archimedean and Aristotelian Elements," *Actes du XIIe Congrès International d'Histoire des Sciences: Paris 1968*, vol. 4 (Blanchard, 1971).

7. Biagioli, *Galileo Courtier*, 159–244.

8. Galileo Galilei, *Discorso al Serenissimo Don Cosimo II Gran Duca di Toscana intorno alle cose che stanno in sull'acqua o in quello si muovono, di Galileo Galilei Filosofo, e Matematico della Medesima Altezza Serenissima* (Florence, 1612), reproduced in *GO* 4.58–141.

9. Several letters by Cigoli, Cesi, and Nozzolini refer to the four authors as "Pippioni." Camerota, *Galileo Galilei*, 218.

10. *Diversi frammenti*, *GO* 4.34.

11. Biagioli, *Galileo Courtier*, 170–72, 177n68 (Biagioli's endnote discusses in more detail Delle Colombe and his failed attempts to elicit a personal response from Galileo).

12. On Nori and Arrighetti, see Biagioli, *Galileo Courtier*, 175–77, esp. 175n58. It is possible that Filippo Salviati originally performed a similar function during the first installment of the dispute but was (regardless of his noble status) not thought to be objective enough given his close friendship with Galileo.

13. The clearest reconstruction of these events can be found in Camerota, *Galileo Galilei*, chap. 5, sec. 3.

14. *Diversi frammenti*, *GO* 4.34.

15. Biagioli, *Galileo Courtier*, 183.

16. Ludovico delle Colombe, *Discorso Apologetico di Lodovico delle Colombe, d'Intorno al Discorso di Galileo Galilei* (Florence, 1612), as it appears in *GO* 4.313–69, 319: "presente l'Illustrissimo ed Eccellentissimo Sig. D. Giovanni Medici, con una nobil brigata di letterati, per sentirci disputare insieme: ma nè si potette far venire a disputa il Sig. Galileo, nè volle far l'esperienza in conveniente grandezza di figura e quantità di

materia; e più tosto si risolvette (giudichi ogn'uno della cagione a suo modo) a mandar in luce un suo trattato intorno a questa materia."

17. Delle Colombe, *Discorso Apologetico*, *GO* 4.319: "sperando far credere altrui col discorrer, quello che non può far veder col senso; atteso che alterando e aggiugnendo, e levando da i patti e dal vero, si può facilmente con false premesse e supposti cavar la conchiusion vera."

18. Biagioli, *Galileo Courtier*, 180–81.

19. Cohen, *How Modern Science Came into the World*, 404–8.

20. Sari Kivistö, *Vices of Learning: Morality and Knowledge at Early Modern Universities* (Brill, 2014), 102, 120, 187.

21. Ludovico Cigoli to Galileo, 23 August 1611, *GO* 11:176: "vi ricordo a venire una volta sola, et poi levarve . . . da torno, et atendere con quelli che sono già famosi e noti al mondo a concorrere, perchè cotesti ucellacci si vogliono far luogho, non per valore propio, ma per la elezione del rivale. Però protestatevi che per una volta farete buono, ma che poi di grazia badi a fare i fatti suoi; et fatela publicha, et non solo colle semplice pratiche, ma principalmente con le buone teorice, acciò poi non vi possino mordere chome fanno, acciò sia manifesto per sodisfazione et degli amici et del Principe."

22. *Diversi frammenti*, *GO* 4.30–31: "perchè, portando così la natura delle contese, quelli che per loro inavvertenza si inducono a voler sostener il falso, più altamente strepitano, e più ne i luoghi pubblici si fanno sentire, che quelli per i quali parla la verità, la quale, se bene con più tempo, con quiete tranquillamente si svela e si denuda; onde io posso stimare che, sì come per le piazze, ne i templi ed altri luoghi publici, molto più frequenti sono state le voci di quelli che dissentono da quanto io asserisco che de gli altri che sentono meco."

23. Nicole Howard, *Loath to Print: The Reluctant Scientific Author, 1500–1700* (Johns Hopkins University Press, 2022); Johns, *Nature of the Book*.

24. Clark, *Academic Charisma*, 88–89; Kivistö, *Vices of Learning*, 95–105.

25. On the ties between Raffaele and Ludovico delle Colombe, see the works of Luigi Guerrini, discussed in the next section of the chapter.

26. Around 1614–1615, rumors circulated about Coresio's sanity, and later scholars have assumed he died around the time he left the Pisan studio. Recent research suggests that he returned to his home country and lived until around 1660. Francesco de Ceglia, "Giorgio Coresio: Note in merito a un difensore dell'opinione di Aristotele," *Physis* 37, no. 2 (2000): 407–11. Coresio's work, *Operetta intorno al galleggiare de corpi solidi* (Florence, 1612), is reproduced in *GO* 4.197–244. On Di Grazia, see Francesco de Ceglia, *De Natantibus: Una disputa ai confini tra filosofia*

e matematica nella Toscana medicea (1611–1615) (Edizioni Giuseppe Laterza, 1999), chap. 5. For a chronological overview of the professors of the Pisan studio, see Danilo Barsanti, "I docenti e le cattedra dal 1543 al 1737," in *Storia dell'università di Pisa, 1343–1737*, ed. Commissione rettorale dell'Università di Pisa, 2 vols. (Pacini Editore, 2000), vol. 1, pt. 2.

27. Vincenzo di Grazia, *Considerazioni di M. Vincenzio di Grazia Sopra'l Discorso di Galileo Galilei intorno alle cose che stanno in su l'acqua, e che in quella si muovono* (Florence, 1613), in *GO* 4.371–440.

28. Accademico Incognito, *Considerazioni sopra il discorso del Sig. Galileo Galilei Intorno alle cose che stanno in su l'acqua o che in quella si muovono, Dedicate alla serenissima D. Maria Maddalena, Archiducessa d'Austria, Gran Duchessa di Toscana: Fatte a difesa, e dichiarazione, dell'opinione d'Aristotile da Accademico Incognito* (Florence, 1612), in *GO* 4.143–96.

29. Biagioli, *Galileo Courtier*, 231; Stillman Drake, *Galileo Studies: Personality, Tradition, and Revolution* (University of Michigan Press, 1970), 170–72; Michele Camerota, "Per segno e per uffizio di animo pronto e leale: Il trattato antigalileiano dell'Accademico Incognito," in *Annali della Facoltà di scienze della formazione dell'Università di Cagliari*, vol. 23 (Grafica del Parteolla, 2000); William R. Shea, "The *Discorso intorno alle cose che stanno in su l'acqua o che in quella si muovono*," unpublished essay, available at the library of the Museo Galileo, where I consulted it in 2018, https://opac.museogalileo.it/imss/resource?uri=341781&found=1&l=it.

30. Accademico Incognito, *Considerazioni*, *GO* 4.147: "Fu impugnato Aristotile nel Discorso del Sig. Galileo Galilei: al quale da certe Considerazioni d'autore per ancora incognito essendosi in buona parte latinamente risposto, molti mi hanno fatta forte istanzia di mandarle in luce, tradotte nel nostro idioma, quasi che ufizio fosse di Proveditore Generale di questo Studio di Pisa publicare le difese d'altri intorno a quella dottrina che qua si professa, e da eccellentissimi filosofi, a ciò condotti e provisionati, s'insegna."

31. James H. Johnson, *Venice Incognito: Masks in the Serene Republic* (University of California Press, 2011), 97–111, 133. On the use of masks by Galileo and his friends, see Wilding, *Galileo's Idol*, chap. 7.

32. Accademico Incognito, *Considerazioni*, *GO* 4.149: "Io, per la fama dell'uomo e dell'esperienze e osservazioni sue, mi posi a leggerlo con molto desiderio."

33. Kivistö, *Vices of Learning*, 76–81; Delle Colombe, *Discorso Apologetico*, *GO* 4.317: "Nondimeno biasimevoli non sono, e giovamento non piccolo n'apportano."

34. Delle Colombe, *Discorso Apologetico*, *GO* 4.317: "Perchè le cose nuove fanno i lor ritrovatori di sì gloriosa memoria, che sono, io non

dirò ammirati solamente, ma reputati come Dei, di qui è che, essendo a pochissimi conceduto questo particolar talento, molti, bramosi di correr cotale arringo, per la mala agevolezza dell'impresa non conseguiscono il desiderato fine d'intorno al vero."

35. Kivistö, *Vices of Learning*, 89–93, 100–101, 187, 227–28.

36. Delle Colombe, *Discorso Apologetico*, *GO* 4.317: "Ma che si trovino intelleti che, a somiglianza di costoro, sperino far nuove apparir le medesime cose di già tralasciate per la falsità loro, in derision degli stessi inventori, e che voglino oggi, che risplende sì bel giorno di verità, far buio altrui con le tenebre dell' intelletto loro. . . . Vorranno costoro contro i primi scrittori del mondo nel pari giostrar, senza sapere di che tempra sien l'armi degli avversari, e senza aver arrotate le sue?"

37. Coresio's *Operetta* (*GO* 4.204), on the other hand, frames Galileo's work as an intellectual exercise and states that he must have published his work to "awaken the minds of scholars, not to give his opinion." De Ceglia, "Giorgio Coresio," 412.

38. Delle Colombe, *Discorso Apologetico*, *GO* 4.317, 327, 333, 355.

39. Polemics were often framed in the language of jousts or sports games, as well as in the more serious language of war. Marian Füssel, "Die Gelehrtenrepublik im Kriegszustand: Zur bellizitären Metaphorik von gelehrten Streitkulturen der Frühen Neuzeit," in *Gelehrte Polemik: Intellektuelle Konfliktverschäftungen um 1700*, ed. Kai Bremer and Carlos Spoerhase (Vittorio Klostermann, 2011).

40. Delle Colombe, *Discorso Apologetico*, *GO* 4.317: "Ora, quantunque il Sig. Galileo quasi in tutte le cose mostri di contrariare ad Aristotile, nel quale è la somma delle filosofiche verità, rinovando molte delle antiche opinioni, non credo già che egli debba annoverarsi tra quegli, stimando io che egli il faccia solo per esercizio di ingegno. Imperochè, se altrimenti fosse, avvenga che per molti suoi meriti e ragioni io il reverisca e reverirò sempre, parendomi che a torto sia doventato un Antiperipatetico, in questo particolare io vorrei poter doventare un Antigalileo, per gratitudine di quel gran principe di tante accademie, capo di tante scuole."

41. Ann Blair, "Natural Philosophy," in *The Cambridge History of Science*, vol. 3, *Early Modern Science*, ed. Katherine Park and Lorraine Daston (Cambridge University Press, 2006), 391; Ann Blair, "Tradition and Innovation in Early Modern Natural Philosophy: Jean Bodin and Jean-Cécile Frey," *Perspectives on Science* 2, no. 4 (1994): 428–54.

42. Daniel Garber, "Descartes among the Novatores," *Res Philosophica* 92, no. 1 (2015): 4–9; Daniel Garber, "Telesio among the Novatores: Telesio's Reception in the Seventeenth Century," in *Early Modern Philosophers and the Renaissance Legacy*, ed. Cecilia Muratori and Gianni Paganini (Springer, 2016); Daniel Garber, "Historicizing Novelty," in

What Reason Promises: Essays on Reason, Nature and History, ed. Wendy Doniger et al. (De Gruyter, 2016).

43. Rising to fame too rapidly (*per saltum* rather than *per gradus*) was rather suspect and could easily lead to accusations of ambition. Marian Füssel, "On the Means of Becoming Famous in the Learned World," in *Scholars in Action: The Practice of Knowledge and the Figure of the Savant in the 18th Century*, vol. 1, ed. André Holenstein et al. (Brill, 2013), 138; Kivistö, *Vices of Learning*, 77–80.

44. Bucciantini, Camerota, and Giudice, *Galileo's Telescope*, 105–6; Huib Zuidervaart, "The 'True Inventor' of the Telescope: A Survey of 400 Years of Debate," in *The Origins of the Telescope*, ed. Albert Van Helden et al. (KNAW Press, 2010).

45. Accademico Incognito, *Considerazioni*, *GO* 4.177–78: "Nondimeno l'apprezzare in ogni tempo i nemici e non lassar che s'avanzino troppo di animo nè di forze, fu precetto militare molto laudato, massimamente quando sono pronti di lingua, d'ingegno acuti, sottili nell'invenzioni e cupidi di gloria. . . . Chi sa che molti giovani, d'ingegno vivace e curiosi di sapere molte cose, allettati dalla novità della dottrina, non si disviassero incautamente dalla strada piana e sicura della filosofia peripatetica, ad altra nuova, piena di rivolgimenti, e che sotto diverse facce rappresenta tutte le cose dell'universo? . . . Troppo perderebbono di frequenza gli Studi e le scuole publiche, e poco sarebbono ascoltati i grand'insegnatori che hanno Aristotile per guida e per primo maestro." On *novatores* and the fear that they would replace current university professors, see Kivistö, *Vices of Learning*, 165, 205–6.

46. Biagioli, *Galileo Courtier*, 230.

47. *GO* 19.487–90, esp. 488: "Secondariamente, conduconsi lettori delle scienze et arti non solamente per la particulare utilità degli scolari privati che a quelle attendono, ma ancora per reputazione et honorevolezza di esse Università si cerca di havere i più insigni e famosi professori di quelle. Non si dubita punto che il Sig.r Galileo si sia talmente avanzato di nome e fama in queste scienze, che forse nessun altro all'età nostra gli metta il piede innanzi." Also, *GO* 19.489: "che quel danaro serva in utile e servizio di quello Studio: del quale che maggior servizio può esser di quello onde gli viene splendore e reputazione? E se ciò gli venga apportato dal S.r Galileo, lascio giudicare a chi sa gli honori che egli ha riceuti e riceve da' primi principi del mondo e da tutti i letterati famosi di Europa, che l'hanno celebrato con i loro scritti, la cui gloria nessuno mi negherà che non redondi in illustrazione dello Studio di Pisa, poi che il Sig.r Galileo si intitola suo primario Matematico." The enrollment numbers of the studio do not suggest Galileo's appointment led to a surge in admissions; see Giuliana Volpi Rosselli, "Il corpo studentesco, i collegi

e le accademie," in *Storia dell'Università di Pisa*, vol. 1, pt. 1 (Pacini Editore, 1993).

48. Tolomeo Nozzolini to Alessandro Marzimedici, 22 September 1612, *GO* 9.289: "E parmi che la lega e l'Incognito procedino contra di lui con inganucci, e non faccino a buona guerra." On Nozzolini's letter, see De Ceglia, *De Natantibus*, 141–56. The archbishop and his relation to Ludovico delle Colombe and his brother Raffaele have been discussed in Luigi Guerrini, "The Archbishop and Astronomy: Alessandro Marzimedici and the 1604 Supernova," in *Copernicus Banned: The Entangled Matter of the Anti-Copernican Decree of 1616*, ed. Natacha Fabri and Federica Favino (Leo S. Olschki, 2018).

49. Galileo of course dedicated his work to the Grand Duke of Tuscany, the Accademico Incognito to the Grand Duchess Maria Maddalena, Delle Colombe to Giovanni de' Medici, Coresio to Don Francesco de' Medici, and Di Grazia to Don Carlo de' Medici.

50. Galileo sent his work out to some twenty recipients, most of them in Italy (in the Veneto, Bologna, and especially Rome), but he also sent copies to Brussels and Prague. Shea, "*Discorso intorno alle cose*," 3n13. The *Sidereus Nuncius* is the only other book that was reprinted in the same time frame, but it appeared only in a pirated edition after its first print run had sold out. Johns, *Nature of the Book*, 25.

51. The authority on Raffaele delle Colombe and the links between him and his brother is Luigi Guerrini. His relevant works include "Echoes from the Pulpit"; *Galileo e la polemica anticopernicana a Firenze* (Polistampa, 2009); *Cosmologie in lotta: Le origini del processo di Galileo* (Polistampa, 2010); "Raffaello delle Colombe et les origines de la polémique anti-galiléenne à Florence, 1610–1615," in *Il processo a Galileo Galilei e la questione galileiana*, ed. Gian Mario Bravo and Vincenzo Ferrone (Edizioni di storia e letteratura, 2010); "Galileo e Raffaelo delle Colombe," in *Il caso Galileo: Una rilettura storica, filosofica, teologica; Convegno internazionale di studi, Firenze, 26–30 maggio 2009*, ed. Massimo Bucciantini et al. (Leo S. Olschki, 2011).

52. On the challenges of establishing the relation between oral and printed sermons, see Stefano Dall'Aglio, "'Faithful to the Spoken Word': Sermons from Orality to Writing in Early Modern Italy," *The Italianist* 34, no. 3 (2014): 463–77; and Stefano Dall'Aglio, "Voices under Trial: Inquisition, Abjuration, and Preachers' Orality in Sixteenth-Century Italy," *Renaissance Studies* 31, no. 1 (2017): 25–42.

53. Raffaele delle Colombe, *Delle Prediche sopra tutti gli Evangeli dell'anno, nelle quali con similitudini, metafore, allegorie retoriche, e considerazioni particolari si dichiarano molti luoghi morali della Sacra Scrittura, ma in senso letterale: Di fra Raffaello delle Colombe dell'Ordine de' predicatori;*

Con tre tauole . . . Volume primo (Florence, 1613); Raffaele delle Colombe, *Prediche della Quaresima con esposizioni di Scritture sacre, Varieta di traslazioni, Dottrine morali, Diuerse erudizioni, e Similitudini, di fra Raffaello Delle Colombe Domenicano, Con Tauole copiose; Delle Scritture esposte, delle cose notabili, e De' titoli di ciascuna Predica . . . 2: Delle prediche di tutto l'anno* (1615; Florence, 1622); Raffaele delle Colombe, *Dupplicato avvento di prediche. Il primo à persone religiose. Il secondo comunemente a tutti. Del padre fra Raffaello delle Colombe domenicano. Con tre tavole copiose . . . Quarto tomo delle sue prediche* (Florence, 1627). The 1615 volume was reprinted in 1622; in what follows I cite the later edition as there appear to be no differences between the two and the 1622 edition is more easily accessible. Only two out of the five sermons are dated. For the others, we have to rely on an *ante quam* date based on the *licenze* in each volume. These can sometimes be supplemented and cross-referenced with specific saints' days to which Delle Colombe refers. Guerrini, *Galileo e la polemica anticopernicana*, 43–45, 65–70; Thomas Mayer, *The Roman Inquisition: Trying Galileo* (University of Pennsylvania Press, 2015), 10, 236–37n21 (hereafter cited as Mayer, *Trying Galileo*, as the author has written several works with the main title *The Roman Inquisition*).

54. Muir, *Culture Wars of the Late Renaissance*, 37–38.

55. Bernadette Paton, "'Una Città Fatticosa': Dominican Preaching and the Defence of the Republic in Late Medieval Siena," in *City and Countryside in Late Medieval and Renaissance Italy: Essays Presented to Philip Jones*, ed. Trevor Dean and Chris Wickham (Bloomsbury, 1990), 109. On the Dominicans' relation to Galileo, see Francesco Beretta, "Les dominicains et le procès de Galilée, ou de l'Inquisition comme instrument de promotion sociale et d'hégémonie intellectuelle," in *I Domenicani e l'Inquisizione Romana: Atti del 3. Seminario internazionale di studi su i domenicani e l'Inquisizione; Roma, 15–18 febbraio*, ed. Carlo Longo (Istituto Storico Domenicano, 2008), 483–98.

56. Arnold Hunt, *The Art of Hearing: English Preachers and Their Audiences, 1590–1640* (Cambridge University Press, 2010); Corrie E. Norman, "The Social History of Preaching: Italy," in *Preachers and People in the Reformations and Early Modern Period*, ed. Larissa Taylor (Brill, 2001); Claude Brémond, Jacques Le Goff, and Jean-Claude Schmitt, *L'"exemplum"* (Brepols, 1982).

57. The sermon appears in the 1613 volume of sermons, which includes *licenze* from 29 April 1609 and 18 June 1610. Guerrini, *Galileo e la polemica*, 53, 64–65.

58. Delle Colombe, *Delle Prediche sopra tutti gli Evangeli dell'anno*, 52: "Ogni cosa quaggiù è pien di bianchissima neve, cioè di beni sensibili, e di gloria terrena, di cui è simbolo la neve" and "e se il voler troppo

curiosamente investigar la gloria celeste è un offuscarsi . . . *qui scrutator est maiestatis opprimeretur à gloria*." The italicized portion cites Proverbs 25:27 (Biblia Sacra Vulgata), which in full reads, "sicut qui mel multum comedit non est ei bonum sic qui scrutator est maiestatis oppremitur a gloria."

59. Delle Colombe, *Delle Prediche sopra tutti gli Evangeli dell'anno*, 63: "Regina prosontuosissima è la Superbia, sostentata dalle spalle di sette Principi, cioè sette peccati mortali, non già che ogni peccato mortale sia superbia, secondo l'affetto, ma la contiene secondo l'effetto, perche chi non obedisce al comandamento di Dio vuole scuoter dal collo il giogo dell'obedienza."

60. Delle Colombe, *Delle Prediche sopra tutti gli Evangeli dell'anno*, 61–62: "Così se chi beve il vino della scienza del mondo, se non vi mette dell'acqua di cui è scritto, *Aqua sapientiae salutaris potabit illum*, darà nel delirio e farà di pazzie. . . . In tal maniera per la superbia ha perturbata la vista, che né anche le cose vicine e molto facili può intendere. Tutte le creature gridano un Dio e i savi del mondo ne posero le centinaia e che cosa più vicina della terra? Qual cosa più sensata che il veder che *Deus firmavit orbem terrae qui non commovebitur*, e con tutto ciò i Copernici dicono che la terra si muove e il cielo sta fermo, perché il sole è il centro della terra, per la qual cosa si può dir di questi che habbian le vertigini, *Dominus miscuit in medio eius spiritum vertiginis et errare fecerunt sicut errat ebrius et vomens*. Sesto se bene la scienza col calor della cognizione dovrebbe esser motivo a riconoscere Iddio, non dimeno per la loro superbia abusandola si raffredano, e danno nel tremore, *Admirati sunt commoti sunt tremor apprehendit eos*, per che han paura di perdere la gloria terrena, non la celesta, *Trepidaverunt timore ubi non erat timor*." I have partially followed the English translation in Guerrini, "Echoes from the Pulpit," 383–84.

61. Delle Colombe, *Delle Prediche sopra tutti gli Evangeli dell'anno*, 57: "Mi direte non è cosa da magnanimo desiderar grandi onori? Signori nò: perche Aristotile nel quarto dell'Etica: cerca la verità delle cose; le ricchezze potentati, e altre cose terrene non istima per grandi il magnanimo, dice; nè si cura esser lodato da gli huomini: in guisa che è vero la magnanimità essere circa gli honori, ma à questo senso: perche ella inclina a far cose onorate, e degne di onore, e contraria alla vanagloria, dice san Tommaso, la quale cerca disordinatamente la gloria."

62. Delle Colombe, *Delle Prediche sopra tutti gli Evangeli dell'anno*, 64: "*Non glorietur dives in divitis suis*, non si vanti di potentia, ò dignità, *Non glorietur fortis in fortitudine sua*, non nella forza della virtù, e bontà della vita, *Nec sapiens in sapientia sua*. Ma in che de[v]e gloriarsi? *Glorietur scire & nosse me*, chi co[n]sidera le miserie del povero conosce

Iddio, che è in lui *Beatus qui intelligit super eg. & p.* segue Ieremia *Qui facio misericordiam*, se f[acci]o misericordia a te, falla à me, *Et iudicium, & iustitiam*, altramente giudicherò delle spese in altre cose superflue, è faronne giustizia."

63. The *licenze* for the 1615 volume were issued in July and September 1613.

64. Delle Colombe, *Prediche della Quaresima*, 311: "Io mi son sempre fatto beffa di questi solenni misuratori, sì come anco di quelli che dalla regolata, dicono essi, proporzione delle Sfere, e dalla moltiplicazione e sottrazione argomentano l'Inferno esser largo miglia settemila ottocento sessantacinque" and "Delle misure dell'Inferno di Dante disputa il Galilei in risposta d'Anton Manetti contro il Vellutello." On the lectures, see Heilbron, *Galileo*, chap. 2.

65. Eileen Reeves, "Speaking of Sunspots: Oral Culture in an Early Modern Scientific Exchange," *Configurations* 13, no. 2 (Spring 2005): 208.

66. Delle Colombe, *Prediche della Quaresima*, 101: "Quell'ingegnoso nostro Mattematico Fiorentino si fa beffe di tutti gli antichi, che facevano il Sole nitidissimo e netto da qual si sia minima macchia, onde ne formarono il proverbio *Querere maculam in Sole*, però egli con lo stromento detto da lui Telescopio fa vedere che ha le sue macchie regolari, come per osservazion di giorni e mesi ha dimostrato. Ma questo farà più veramente Iddio, perché *Coeli non sunt mundi in conspectu eius*, se si troveranno le macchie ne' soli dei giusti, pensate voi, se si troveranno nelle lune degli instabili peccatori." A note in the margins referred readers to Galileo's work on the sunspots: "Galileus in de Maculis Solis." Guerrini, *Galileo e la polemica*, 58, argues this sermon must have been delivered during Lent 1614, as the first Sunday of Lent in 1613 fell on 24 February, and Galileo's work on the sunspots was not published until March that year. Yet, like so many of his letters, the letters on which Galileo's *Istoria e dimostrazioni* were based may have circulated before the book was published.

67. I have not found any ancient usage of this expression, but that the proverb had this specific meaning in early modern Europe is demonstrated in a 1641 work whose author equates this expression with another one ("nodum in scirpo quaeri"—"seeking a knot in a bulrush"), which can be found in the work of the Roman playwright Terence. Petrus Servius, *Dissertatio Philologica de Odoribus* (Rome, 1641) 13; Terence, *The Woman of Andros. The Self-Tormentor. The Eunuch*, ed. and trans. John Barsby (Harvard University Press, 2001), 161–62n52.

68. Eileen Reeves, *Painting the Heavens: Art and Science in the Age of Galileo* (Princeton University Press, 1997), 139–48.

69. Steven F. Ostrow, "Cigoli's Immacolata and Galileo's Moon: As-

tronomy and the Virgin in Early Seicento Rome," *Art Bulletin* 78, no. 2 (June 1996): 231–32.

70. Job 15:4 (New International Version).

71. It is worth pointing out that Delle Colombe's last line—"the moons of fickle sinners"—invokes the moon as symbol of fickleness and instability, while it was more often, and even in a later sermon by Delle Colombe, taken as a symbol of purity. On these conflicting views of the moon, see Ostrow, "Cigoli's Immacolata," 233–34; and Reeves, *Painting the Heavens*, 139–48, 167.

72. Delle Colombe, *Dupplicato avvento di prediche*, 205: "Secondo Avvento Predicato dall'Autore nel Duomo di Fiorenza. L'anno 1615. Nella Prima Domenica dell'Avvento. *Erunt signa in Sole, & Luna, & Stellis. Luc. 21.*"

73. Delle Colombe, *Dupplicato avvento di prediche*, 212: "Ma vuoi vedere la malignità del peccato, che appesta insino i cieli e fa nero il Sole e oscura le Stelle? Lo disse San Paolo Apostolo, *Sine sanguinis effusione non sit remissio. Necesse est ergo exempla quidem Coelestium his mundari melioribus hostiis.* La scrittura è difficile, e io con San Tommaso d'Aquino intendo il Cielo in senso proprio, e non metaforico. Ma come può il peccato macchiare il Cielo? dispregiandolo, volgendosi le spalle, quasi che il Cielo fusse fango." The italicized portion of the passage quoted in the main text is a slightly shorter version of Hebrews 9:22–23 (Biblia Sacra Vulgata).

74. Delle Colombe, *Dupplicato avvento di prediche*, 355–56: "Fu pensiero di Seneca che lo specchio fusse ritrovato per poter contemplare il Sole. Non pareva convenevol cosa che l'uomo non potesse considerar la bellezza della maggior luce che comparisca nel teatro del mondo. . . . Ma come và meglio questo à Maria? Chi potrebbe l'infinita luce del divin sole fisamente risguardare, se non fusse questo Verginale specchio, che in se il concepisce, e al mondo lo rende?" The analogy between the mirror and Mary was not uncommon in early modern Europe; see Reeves, *Painting the Heavens*, 144.

75. Delle Colombe, *Dupplicato avvento di prediche*, 355–56: "Fu pensiero di Seneca che lo specchio fusse ritrovato per poter contemplare il Sole. Non pareva convenevol cosa che l'uomo non potesse considerar la bellezza della maggior luce che comparisca nel teatro del mondo. Ma perché l'occhio mortale per la debolezza della sua vista non può fissare lo sguardo per lo troppo suo gran lume, almeno il rimiri in un chiaro cristallo dentro cui la sua bella immagine il sole vi rappresenti. Ma d'uno che cerchi difetto, anche dove non è non dicevano gli antichi *Querit maculam in Sole*? Il Sole è senza macchia, e la madre del Sole è senza macchia."

76. Eileen Reeves, *Galileo's Glassworks: The Telescope and the Mirror*

(Harvard University Press, 2008). See also Figliucci's poem, discussed in chapter 3 of this book, and those by Iacopo Soldani and Iacopo Cigognini, in Vaccalluzzo, *Galileo Galilei nella poesia*, 19, 31.

77. Delle Colombe, *Discorso Apologetico*, *GO* 4.317: "che sono, io non dirò ammirati solamente, ma reputati come Dei."

78. *GO* 19.299–305; Maurice A. Finocchiaro, *The Galileo Affair: A Documentary History* (University of California Press, 1989), 27, 49–54. By April 1616, the letter had made it to Francis Bacon in England. *GO* 11.255.

79. *GO* 19.297: "essendomi capitato alle mani una scrittura, corrente qua nelle mani di tutti, fatta da questi che domandono Galileisti, affermanti che la terra si muove et il cielo sta fermo, seguendo le posizioni di Copernico, dove, a giud[izio] di tutti questi nostri Padri di questo religiosissimo convento di S. Marco, vi sono dentro molte proposizioni che ci paiono o sospette o temerarie"; and "vedendo non solo che questa scrittura corre per le mani d'ogn'uno, senza che veruno la rattenga de' superiori."

80. *GO* 19.297–98: "Mi protesto ch'io tengo tutti costoro, che si domandono Galileisti, huomini da bene e buon Christiani, ma un poco saccenti e duretti nelle loro opinioni . . . et insomma per fare il bell'ingegno si dicono mille impertinenze e si seminano per tutta la città nostra."

81. Thomas Mayer has pointed out that Lorini framed his letter in such a way that it was up to its addressee, Cardinal Paolo Emilio Sfondrati, to decide whether to interpret it as a formal complaint or a personal note. Mayer, *Trying Galileo*, 20–21. For the report, see *GO* 19.305; Mayer, *Trying Galileo*, 32–33; and Finocchiaro, *Galileo Affair*, 135–36.

82. The inquisitor's correspondence is in *GO* 19.298, 306, 311–12; the letters between Castelli and Galileo are in *GO* 12.153–54, 158–59, 161–62, 165–66. In a letter to Piero Dini (see below), Galileo suggested that Lorini might have tinkered with his words. This, as well as the differences between Lorini's copy and the version of the letter Galileo sent to the Inquisition, led later historians to suspect Lorini of having exaggerated the potentially heretical elements of Galileo's letter, befitting Galileo's self-fashioning as a victim of the malicious envy of the Dominican friars; see, for instance, Heilbron, *Galileo*, 208. After Salvatore Ricciardo's recent rediscovery of Galileo's autograph letter to Castelli, it has become clear that the wording of Galileo's original letter to Castelli is exactly the same as Lorini's copy, which means that Galileo sent in a different and toned-down version to the Inquisition. Michele Camerota, Franco Giudice, and Salvatore Ricciardo, "The Reappearance of Galileo's Original Letter to Castelli," *Notes and Records: The Royal Society Journal for the History of Science* 73, no. 1 (March 2019): 11–28.

83. Mayer, *Trying Galileo*, 20–21.

84. Caccini's opening statement, at *GO* 19.307, reads as follows: "Presi per tanto occasione da questo luogo, da me prima in senso litterale et poi in sentimento spirituale, per salute delle anime, interpretato, di riprovare, con quella modestia che conviene all'offitio che tenevo, una certa opinione già di Nicolò Copernico, et in questi tempi, per quel ch'è ublichissima fama nella città di Firenze, tenuta et insegnata, per quanto dicono, dal Sig.r Galileo Galilei matematico."

85. *GO* 19.308–11: "un certo gentil'huomo Fiorentino degl' Attavanti." The three propositions are "God is not otherwise a substance, but an accident'; "God is sensuous because there are in him divine senses"; and "in truth the miracles said to have been by the saints are not real miracles." Translation based on Finocchiaro, *Galileo Affair*, 137–38.

86. This preacher has never been identified, despite several attempts. Mayer, *Trying Galileo*, 245n96. Massimo Bucciantini finds a parallel with Giordano Bruno's 1600 case in the accusation against Galileo's students; see Bucciantini's *Contro Galileo: Alle origini dell'affaire* (Leo S. Olschki, 1995), 40–47.

87. *GO* 19.309–10: "Da molti è tenuto buon Cattolico; da altri è tenuto per sospetto nelle cose della Fede, perchè diconò sii molto intimo di quell Fra Paolo Servita, tanto famoso in Venetia per le sue impietà, et dicono che anco di presente passino lettere tra di loro."

88. Massimo Rospocher and Rosa Salzberg, "An Evanescent Public Sphere: Voices, Spaces and Publics in Venice during the Italian Wars," in *Beyond the Public Sphere: Opinions, Publics, Spaces in Early Modern Europe*, ed. Massimo Rospocher (Il Mulino/Duncker & Humblot, 2012), 99; Stephen J. Millner, "The Florentine Piazza della Signoria as Practiced Place," in *Renaissance Florence: A Social History*, ed. John T. Paoletti and Roger Crum (Cambridge University Press, 2006).

89. Bacon, *Essays or Counsels, Civil and Moral*, 207.

90. Bacon, *Essays or Counsels, Civil and Moral*, 46.

91. Horodowich, "Gossiping Tongue"; Rospocher and Salzberg, "Evanescent Public Sphere."

92. *GO* 19.312: "et quanto prima potrò havere li testimonii prodotti, de'quali alcuni sono hora occupati nelle predicationi quadragesimali, eseguirò subito il contenuto della detta lettera"; *GO* 19.313: "Perchè il P.F. Ferdinando Gimenes dell'ordine de' Predicatori, che intorno al fine di Marzo passato partì da questa città per Milano . . . non mi è parso di cominciare l'essamine delle persone nominate nella denuntia del P.F. Thomasso Caccini, del medesimo ordine, contro Galileo Galilei, come già scrissi a V.S.Ill.ma et R.ma, ma di aspettare et vedere prima le depositioni di detto P. Gimenes intorno alle tre propositioni che si pretendono asser-

te dalli discepoli di detto Galileo, che è fondamento principale di quanto si possa pretendere contro detto Galileo et che solo ha bisogno di prova."

93. *GO* 19.314–20.

94. *GO* 19.316–17: "dico bene che, conforme quello ch'ho sentito dire dell'opinione del moto della terra et fermezza del cielo, et anco a quello ch'ho sentito dire da quelli che conversano seco, dico esser doctrina contraposita *ex diamatro* alla vera theologia et filosofia. . . . Ho sentito alcuni suoi scolari, i quali hanno detto che la terra si muove et che il cielo è immobile. . . . Io non credo che il detto Piovano Attavanti assertivamente dicesse et credesse le sopradette cose, perchè mi pare che lui stesso dicesse che si rimettava alla Chiesa, et che il tutto dicesse *disputationis gratia*."

95. *GO* 19.318–19.

96. *GO* 19.320.

97. Copernicus's *De Revolutionibus* was not banned in its entirety, but parts of it were to be censured; on the ban and the different ways in which it could be interpreted, see Rivka Feldhay, *Galileo and the Church: Political Inquisition or Critical Dialogue?* (Cambridge University Press, 1995). For an English translation of the injunction, see Finocchiaro, *Galileo Affair*, 147.

98. Galileo to Dini, 16 February 1615, *GO* 5.291–95; Feingold, "Grounds for Conflict."

99. *GO* 5.305–48. For a discussion of the letter, see Ernan McMullin, "Galileo on Science and Scripture," in *The Cambridge Companion to Galileo*, ed. Peter Machamer (Cambridge University Press, 1998); and Richard S. Westfall, *Essays on the Trial of Galileo* (Vatican Observatory Publications, 1981), 17–21.

100. Galileo to Dini, 16 February 1615, *GO* 5.291–95.

101. Dini to Galileo, 7 March 1615, *GO* 12.151.

102. To be precise, Foscarini sent Bellarmine his book (written in the form of a letter) as well as a letter containing his response to the Inquisition's judgment of his book (they considered it rash in three ways). Richard Blackwell, *Galileo, Bellarmine, and the Bible: Including a Translation of Foscarini's Letter on the Motion of the Earth* (University of Notre Dame Press, 1992), 253–54; Dietz Moss, *Novelties in the Heavens*, chaps. 5 and 6.

103. Bellarmine to Foscarini, 12 April 1615, *GO* 12.171: "P.o Dico che mi pare che V. P. et il Sig.r Galileo facciano prudentemente a contentarsi di parlare ex supposizione e non assolutamente, come io ho sempre creduto che habbia parlato il Copernico."

104. Bellarmine to Foscarini, 12 April 1615, *GO* 12.172: "3o. Dico che quando ci fusse vera demostratione che il sole stia nel centro del

mondo e la terra nel 3° cielo, e che il sole non circonda la terra, ma la terra circonda il sole, allhora bisogneria andar con molta consideratione in esplicare le Scritture che paiono contrarie, e più tosto dire che non l'intendiamo, che dire che sia falso quello che si dimostra. Ma io non crederò che ci sia tal dimostratione, fin che non mi sia mostrata: nè è l'istesso dimostrare che supposto ch'il sole stia nel centro e la terra nel cielo, si salvino le apparenze, e dimostrare che in verità il sole stia nel centro e la terra nel cielo; perchè la prima dimostratione credo che ci possa essere, ma della 2.a ho grandissimo dubbio, et in caso di dubbio non si dee lasciare la Scrittura Santa, esposta da' Santi Padri."

105. McMullin has taken the first position, while Feldhay and Cohen both interpret Bellarmine's third point as at least allowing for the possibility that proof might be found. McMullin, "Galileo on Science and Scripture," 279–83, 332n36; Feldhay, *Galileo and the Church*, 35–36; Cohen, *How Modern Science Came into the World*, 421.

106. He most likely obtained the letter from Dini in April 1615. *GO* 11.173.

107. McMullin, "Galileo on Science and Scripture," 287. For a discussion of Galileo's rhetorical strategies in this letter, see Jean Dietz Moss, "Galileo's Letter to Christina: Some Rhetorical Considerations," *Renaissance Quarterly* 36, no. 4 (Winter 1983): 547–76; and Howard, *Loath to Print*, 25.

108. The extent of the letter's circulation in 1615 is unclear. McMullin thinks it unlikely Bellarmine had a copy; Westfall disagrees, arguing that even if Bellarmine had not seen (or heard of) the work before December 1615, Galileo most likely brought a copy with him on his visit to the city. McMullin, "Galileo on Science and Scripture," 289; Westfall, *Essays on the Trial of Galileo*, 17–20.

109. Camerota, *Galileo Galilei*, 274–78.

110. The warning had been delivered by the Florentine ambassador in Rome to the grand duke's secretary. Piero Guicciardini to Curzio Picchena, 15 December 1615, *GO* 12.207: "questo non è paese da venire a disputare della luna, né da volere, nel secolo che corre, sostenere né portarci dottrine nuove." See Cohen, *How Modern Science Came into the World*, 190, 422. On Paul V, his Inquisition policies, and Orsini, see Mayer, *Trying Galileo*, 37–40, 46.

111. Galileo to Curzio Picchena, 6 February 1616, *GO* 12.230: "Ma perchè alla causa mia viene annesso un capo che concerne non più alla persona mia che all'università di tutti quelli che da 80 anni in qua, o con opere stampate o con scritture private o con ragionamenti pubblici e predicazioni o anco in discorsi particolari, havessero aderito o aderissero a certa dottrina et opinione non ignota a V. S. Ill.ma, sopra la determi-

nazione della quale hora si va discorrendo per poterne deliberare quello che sarà giusto et ottimo; io, come quello che posso per avventura esserci di qualche aiuto per quella parte che depende dalla cognizione della verità che ci vien summnistrata dalle scienze professate da me, non posso nè devo trascurare quell'aiuto che dalla mia coscienza, come cristiano zelante e cattolico, mi vien sumministrato."

112. Piero Guicciardini to Curzio Picchena, 4 March 1616, *GO* 12.241–42: "ma se voleva tenere questa openione, tenerla quietamente, senza far tanto sforzo di disporre e tirar gl'altri a tener l'istesso."

113. Translation from Finocchiaro, *Galileo Affair*, 147.

114. Translation from Finocchiaro, *Galileo Affair*, 147.

115. Galileo to Curzio Picchena, 6 March 1616, *GO* 12:243–45; Galileo to Curzio Picchena, 12 March, *GO* 12.247–48.

116. Curzio Picchena to Galileo, 20 March 1616, *GO* 12.250: "Hanno avuto molto contento di sentire che ella havesse havuto da S.S.tà così benigna audienza; et parendo loro che V.S. habbia hora la sua riputazione in tutti i conti, m'hanno comandato di esortarla per parte loro che si quieti et non tratti più di coteste materie, et più tosto se ne torni. V.S. sa che l'Alt.e loro l'amano, et le dicono questo per suo bene et per sua quiete."

117. Curzio Picchena to Galileo, 23 May 1616, *GO* 12.261.

118. On the rumors, see the letters by Benedetto Castelli from Pisa (20 April 1616, *GO* 12.254) and from Gianfrancesco Sagredo from Venice (23 April 1616, *GO* 12.257–59). See also Finocchiaro, *Galileo Affair*, 31, 153. For the original text of the certificate, see *GO* 19.342, 348.

5. GLORY OR INFAMY

1. Evangelista Torricelli, *Lezioni accademiche di Evangelista Torricelli* (Florence, 1715), 55. I am grateful to J. B. Shank for calling my attention to this intriguing lecture, which is part of a series Torricelli delivered before the Accademia della Crusca and the Accademia Fiorentina between 1642 and 1647 and published in 1715. Two manuscript versions are extant, both held in the Biblioteca Nazionale Centrale di Firenze (BNCF), Gal. 133 IV, Discepoli 23, Torricelli Evangelista 3, Opere letterarie, 1r–10v and 23r–30v. The second version is a much neater version of the first. The small differences between the manuscript versions and the printed text do not seem to be significant to Torricelli's argument.

2. Stefano Gattei, "From Banned Mortal Remains to the Worshipped Relics of a Martyr of Science: Vincenzo Viviani and the Birth of Galileo's Mythography," in *Brains and Remains of Scientists*, ed. Marco Beretta et al. (Science History Publications, 2016), esp. 67–69.

3. For a discussion of Torricelli's relation with the Accademia della Crusca and the publication of his lecture series, see Domenico De Martino, ed., *Lezioni accademiche d'Evangelista Torricelli* (Biblion, 2009).

4. Torricelli, *Lezioni accademiche*, 55.

5. The academy's diarist, Benedetto Buonmattei, noted on 13 August 1643 that the lecture had drawn many counterarguments, but he did not offer any specifics—except for a note stating that the academy member Filippo Galilei would reply to the lecture during their next session. Archivio dell'Accademia della Crusca (AAC), Fascetta 76: *Diario del Ripieno*, 43v: "Oggi l'inno[minato] Torricelli fece una dotta lezione paradossica della Fama, contro al parer d'Aristotele. Gli fu da molti dato contro. Fu commesso all'Innom[inato] Galilei che risponda nella seguente."

6. On the role of the different religious orders and the role played by the 1616 injunction, see Blackwell, *Galileo, Bellarmine, and the Bible*; Richard Blackwell, *Behind the Scenes at Galileo's Trial* (University of Notre Dame Press, 2008); and Feldhay, *Galileo and the Church*. Comprehensive entry points into the trial in general are Finocchiaro's *Galileo Affair* and *Retrying Galileo*.

7. See especially Francesco Beretta, "Galilée devant le tribunal de l'Inquisition: Une relecture des sources" (Université de Fribourg [dissertation], 1998); and Mayer, *Trying Galileo*.

8. Within the Catholic Church, various institutions were responsible for dealing with heretics and the spiritual health of the Catholic community. Most important for the context of this chapter are the Congregation of the Index of Prohibited Books, the Master of the Sacred Palace, and the Roman Inquisition or the Holy Office. The focus of this chapter is, for the most part, on the last. For a discussion of the specific responsibilities of the Roman Inquisition and the changes it underwent during different papacies, see Thomas Mayer, *The Roman Inquisition: A Papal Bureaucracy and Its Laws in the Age of Galileo* (University of Pennsylvania Press, 2013).

9. The events of 1632 and 1633 can be reconstructed from Inquisition records and (diplomatic) correspondence. The former can be found in *GO* 19.272–421 and the latter in *GO* 14. Translations of the most important documents can be found in Finocchiaro, *Galileo Affair*, which also includes a chronologic overview (297–308).

10. Finocchiaro, *Galileo Affair*, 29.

11. *GO* 19.322–23.

12. For a concise discussion of the precise events and different documents relating to this ban, see Heilbron, *Galileo*, 218–21. For an excellent discussion of how early modern scholars and censors approached the concept of utility, see Hannah Marcus, *Forbidden Knowledge: Medi-*

cine, Science, and Censorship in Early Modern Italy (University of Chicago Press, 2020). The epilogue to that work (226–36) discusses the ban on Copernicus and Galileo's subsequent actions from this perspective.

13. Finocchiaro, *Galileo Affair*, 30–31.

14. On Galileo's visit to Rome, see Shea and Artigas, *Galileo in Rome*, 94–122.

15. *GO* 13.182; Finocchiaro, *Galileo Affair*, 303.

16. Shea and Artigas, *Galileo in Rome*, chap. 5 and 161–62. On the *Dialogo*'s print run, see Galileo to Cesare Marsigli, 5 July 1631, *GO* 14.281.

17. *GO* 19.348–56.

18. Finocchiaro, *Galileo Affair*, 223.

19. Biagioli, *Galileo Courtier*, 335–36.

20. *GO* 19.321–22; translation from Finocchiaro, *Galileo Affair*, 147.

21. *GO* 19.348.

22. *GO* 19.336–42. Galileo's defense actually corresponds with a second document in the Inquisitorial archives, namely Bellarmine's 26 February report of his delivery of the injunction. The report does not say that Bellarmine told Galileo not to *discuss* the doctrine in any way, but it does state that Bellarmine had told Galileo, in the presence of several witnesses, "to abandon completely the abovementioned opinion that the sun stands still at the center of the world and the earth moves, and henceforth not to hold, teach, or defend it in any way whatever, either orally or in writing." Finocchiaro, *Galileo Affair*, 147.

23. Vincenzo Maculano da Firenzuola to Cardinal Francesco Barberini, 28 April 1633, *GO* 15.106: "Finalmente proposi io un partito, che la S. Congregatione concedesse a me la facoltà di trattare estraiudicialmente col Galileo . . . e si dispose a confessarlo giuditialmente: mi dimandò però alquanto di tempo per pensare al modo co 'l quale egli poteva honestare la confessione, chè quanto alla sostanza spero seguirà nella maniera sodetta."

24. Vincenzo Maculano da Firenzuola to Cardinal Francesco Barberini, 28 April 1633, *GO* 15.107: "Che in questo modo si ponga la causa in termine che senza difficoltà si possi spedire. Il Tribunale sarà nella sua reputatione, co'l reo si potrà usare benignità . . . e ciò fatto, si potrà habilitare alla casa per carcere, come accennò V. E."

25. Galileo's deposition, 30 April 1633, *GO* 19.342–43: "avidior sim gloria quam satis sit . . . è stato dunque l'error mio, e lo confesso, di una vana ambitione e di una pura ignoranza et inavertenza." Also see Finocchiaro, *Galileo Affair*, 278.

26. Translation from Finocchiaro, *Retrying Galileo*, 16–17.

27. Formal heresy was the most serious crime, followed by suspected

heresy, which was in turn divided into three categories indicating the seriousness of the offense: violent, vehement, and slight. Finocchiaro, *Retrying Galileo*, 11–12; Ugo Baldini and Leen Spruit, eds., *Catholic Church and Modern Science: Documents from the Archives of the Roman Congregations of the Holy Office and the Index*, vol. 1, *Sixteenth-Century Documents*, book 1 (Libreria Editrice Vaticana, 2009), 41–44.

28. F. Beretta, "Galilée devant le tribunal d'Inquisition," 28–29.

29. Eliseo Masini, *Sacro arsenale, ouero Prattica dell'Officio della S. Inquisitione, Ampliata* (Genoa, 1625), 368: "E di sì brutta, e di sì horribil nota il delitto d'heresia, che chi lo comette, incorre nell'infamia iuris, & facti; e perciò non si presume così agevolmente, alcuno esser heretico: e chiunque dice, questi, ò quegli esser tale; conviene, che lo provi." On Inquisition manuals as a source, see Baldini and Spruit, *Catholic Church and Modern Science*, 44–47.

30. For this development, see especially Edward Peters, "Wounded Names: The Medieval Doctrine of Infamy," in *Law in Mediaeval Life and Thought*, ed. Edward B. King and Susan J. Ridyard (University of the South Press, 1990); and Peter Landau, *Die Entstehung des Kanonischen Infamiebegriffs von Gratian bis zur Glossa Ordinaria* (Bohlau, 1966).

31. Peters, "Wounded Names," 65–69, 83–86. This second type of infamia could relate to a person's profession (as with prostitutes) or origin (as in the case of a bastard), but it could also be applied to individuals who were rumored to engage in activities deemed immoral. Bowman, "Infamy and Proof in Medieval Spain," 104–10. In those latter cases, infamia facti could lead to an inquisitorial procedure investigating whether the rumors were true.

32. F. Beretta, "Galilée devant le tribunal d'Inquisition," 145.

33. Masini, *Sacro arsenale*, 364: "Dannasi la memoria dell'heretico morto; avvenga che, vivendo, non sia stato diffamato d'heresia." On bearers of cultural memory, see Jan Assmann, *Cultural Memory and Early Civilization: Writing, Remembrance, and Political Imagination* (Cambridge University Press, 2012), 39–40.

34. Masini, *Sacro arsenale*, 279: "condanniamo la memoria di esso, come di formale, e consummata heretico, pertinace, & impenitente; e lo dichiariamo infame, e scommunicato, & indegno d'Ecclesiastica sepoltura: e perciò ordiniamo, che l'ossa di lui, se pure dall'ossa de'fedeli si potranno discernere, siano dissotterate, & portate fuori nel Cimiterio, & in detestatione del suo grave delitto publicamente abbruciate. Di più rilasciamo al braccio secolare la statua del detto N. [Nome] qui presente, accioche essa parimente venga (come di ragione conviene) abbrucciata."

35. Masini, *Sacro arsenale*, 331; F. Beretta, "Galilée devant le tribunal d'Inquisition," 145–46.

36. For simplicity's sake, I use the term *damnatio memoriae* here, even though it was not in use during Antiquity. Useful discussions of the term and its continued use may be found in Charles W. Hedrick, *History and Silence: Purge and Rehabilitation in Late Antiquity* (University of Texas Press, 2000), 93; and Harriet I. Flower, *The Art of Forgetting: Disgrace and Oblivion in Roman Political Culture* (University of North Carolina Press, 2006), xix.

37. For an overview of common forms the *damnatio memoriae* could take, see Hedrick, *History and Silence*, 89–130.

38. Ross Poole, "Enacting Oblivion," *International Journal of Politics, Culture, and Society* 22, no. 2 (2009): 149–57; Judith Pollmann, *Memory in Early Modern Europe, 1500–1800* (Oxford University Press, 2017), 140–58; Hedrick, *History and Silence*, 93.

39. Roman law initially made clear distinctions between the application of *damnatio memoriae*, which was restricted to cases of political treason, and infamy, which followed a wider range of crimes. Infamia was extended as a punishment to cases of heresy in the fourth century CE; see Sarah Bond, "Altering Infamy: Status, Violence, and Civic Exclusion in Late Antiquity," *Classical Antiquity* 33, no. 1 (2014): 2; and A. H. J. Greenidge, *Infamia: Its Place in Roman Public and Private Law* (Oxford, 1894), 148–49. For the early modern application of memory sanctions in political and civic cases, see Tracey E. Robey, "Damnatio Memoriae: The Rebirth of Condemnation of Memory in Renaissance Florence," *Renaissance and Reformation* 36, no. 3 (2013): 5–32; Pollmann, *Memory in Early Modern Europe*, chap. 6; and Poole, "Enacting Oblivion."

40. Masini, *Sacro arsenale*, 331; F. Beretta, "Galilée devant le tribunal d'Inquisition," 145–46, Robey, "Damnatio Memoriae," 16–17.

41. This crime also typically incurred some form of infamia, but its application was not as clearly delineated. Peters, "Wounded Names," 76. Another possibility is that the status of Galileo's fama was intentionally left vague, so as to sow confusion.

42. Heilbron, *Galileo*, 267.

43. Masini, *Sacro arsenale*, 316: "Nel carcerare i Rei bisogna usare grandissima prudenza, perche la sola carceratione per lo delitto d'heresia apporta notabile infamia al carcerato. Onde havra molto bene a considerarsi, e la natura degl'indicii, e la qualità de'testimoni, e conditione del Reo, per caminare cautamente, e sicuramente."

44. Shea and Artigas, *Galileo in Rome*, 179–80.

45. See especially the letters from 14, 16, and 19 February, 27 February, 6 March, 16 April, and 29 May, all in *GO* 15; translations can be found in Finocchiaro, *Galileo Affair*, 242–54.

46. *GO* 19.284. Galileo was only granted permission to go to Siena after negotiations by the Florentine ambassador in Rome; see GO 15.165.

47. *GO* 19.393: "Il Galileo ha seminato in questa città opinioni poco cattoliche, fumentato da questo Arcivescovo suo hospite, quale ha sugerito a molti che costui sia stato ingiustamente agravato da cotesta Sacra Congregatione, e che non poteva né doveva reprobar le opinioni filosofische, da lui con ragioni invincibili mattematiche e vere sostenute, e che è il prim'homo del mondo, e viverà sempre ne'suoi scritti, ancor prohibiti, e che da tutti moderni e migliori vien sequitato. E perché questi semi da bocca d'un prelato potriano produrre frutti perniciosi, se ne dà conto etc."

48. *GO* 19.286.

49. Finocchiaro, *Retrying Galileo*, 57–59.

50. *GO* 19.286–88.

51. Galileo to Michelangelo Buonarroti the Younger, 26 June 1638, *GO* 17.346–47.

52. Finocchiaro, *Retrying Galileo*, 63–64.

53. Masini, *Sacro arsenale*, 237. On the policy and practice of (medical) book burning, see Marcus, *Forbidden Knowledge*, 38–50, 176–79.

54. *GO* 19.283n15.

55. Translation from Finocchiaro, *Retrying Galileo*, 43.

56. On the effectiveness of the prohibition, see Baldini and Spruit, *Catholic Church and Modern Science*, 69–91; Daniel Stolzenberg, "The Holy Office in the Republic of Letters: Roman Censorship, Dutch Atlases, and the European Information Order, circa 1660," *Isis* 110, no. 1 (2019): 10–11; and Marcus, *Forbidden Knowledge*, 234.

57. Galileo to Nicolas-Claude Fabri de Peiresc, 12 May 1635, *GO* app. 2.337: "Da questo, e dall'esser state raccolte in Firenze, et in Roma tutte l'opere mie, sì che più non se ne trovano per le librerie, apertamente si scorge che si fa ogni opera per levar dal mondo la mia memoria."

58. Finocchiaro, *Retrying Galileo*, 60.

59. Finocchiaro, *Retrying Galileo*, 61. On the various international attempts to publish the *Discorso*, see Renée Raphael, "Printing Galileo's *Discorsi*: A Collaborative Affair," *Annals of Science* 69, no. 4 (2012): 483–513.

60. Harald Hendrix, ed., *Writers' Houses and the Making of Memory* (Routledge, 2008); Susan Gaylard, *Hollow Men: Writing, Objects, and Public Image in Renaissance Italy* (Fordham University Press, 2003), 23; Karl A. E. Enenkel and Konrad Adriaan Ottenheym, eds., *The Quest for an Appropriate Past in Literature, Art and Architecture* (Brill, 2018).

61. For correspondence regarding Galileo's death and the possible construction of a funerary tomb, see *GO* 18.378–82. The grand duke reserved 3,000 *scudi* for the monument. Shea and Artigas, *Galileo in Rome*, 199.

62. *GO* 19.559–62. The report is not dated and not signed. An earlier

report, drawn up in 1641, even discussed the validity of Galileo's will (those who had incurred infamia lost the necessary legal status for drafting a will). *GO* 19.535–37.

63. The most thorough discussion of how the monument eventually materialized is in Paolo Galluzzi, "I sepolcri di Galileo: Le spoglie 'vive' di un eroe della scienza," in *Il Pantheon di Santa Croce a Firenze*, ed. L. Berti (Cassa di Risparmio, 1993); and an English version, Paolo Galluzzi, "The Sepulchers of Galileo: The 'Living' Remains of a Hero of Science," in *The Cambridge Companion to Galileo*, ed. Peter Machamer (Cambridge University Press, 2006).

64. Francesco Niccolini to Giovanni Battista Gondi, 25 January 1642, *GO* 18.379: "Voleva dire che non era punto d'esempio al mondo che S. A. facesse questa cosa, mentre egli è stato qui al Santo Offitio per una opinione tanto falsa e tanto erronea, con la quale anche ha impressionati molti altri costà, e dato uno scandalo tanto universale al Cristianesimo con una dottrina stata dannata."

65. Inquisition decree of 23 January 1642, *GO* 19.290: "scandalizentur boni."

66. Francesco Barberini to Giovanni Muzzarelli, 25 January 1642, *GO* 18.379–80.

67. *GO* 19.283.

68. On the need for public atonement (*vendetta*) in cases of heresy, see F. Beretta, "Galilée devant le tribunal d'Inquisition," 28–29, 141, 144. On the spectacle of punishment, see Michel Foucault, *Discipline and Punish: The Birth of the Prison* (Pantheon Books, 1977), 32–72; and on honor and shame as decisive concepts that influenced the penal system in early modern Europe, see Florike Egmond, "Execution, Dissection, Pain, and Infamy—A Morphological Investigation," in *Bodily Extremities: Preoccupations with the Human Body in Early Modern European Culture*, ed. Florike Egmond and Robert Zwijnenberg (Routledge, 2003).

69. Masini, *Sacro arsenale*, 308: "Qualunque non havrà, spontaneamente comparendo, accusato se stesso, ma sarà stato denontiato, o per altro modo giudiciale, secondo l'ordine di ragione, indiciato, inquisito, processato, e colpevole ritrovato d'heresia formale, dovrà, pentendosi, abiurare publicamente con l'habitello. Quelli, che abiurano solo come vehementemente sospetti d'heresia, o d'apostasia, ancorche ciò segua alle volte in publico, non devono però portar l'habitello." A letter by Jean Jacques Bouchard claims Galileo did wear an "abito di penitenza," yet this is the only source mentioning it. *GO* 15.166.

70. Mayer, *Trying Galileo*, 209–10. Finocchiaro instead posits that Galileo abjured in a private setting. Finocchiaro, *Retrying Galileo*, 26.

71. Finocchiaro, *Retrying Galileo*, 6.

72. *GO* 19.283.

73. Translation of Antonio Barberini's letter to the inquisitors and nuncios, 2 July 1633, from Finocchiaro, *Retrying Galileo*, 27.

74. Flower, *Art of Forgetting*, 9–10.

75. Hedrick, *History and Silence*, 91.

76. Karl Härter, "Images of Dishonoured Rebels and Infamous Revolts: Political Crime, Shaming Punishments and Defamation in the Early Modern Pictorial Media," in *Images of Shame: Infamy, Defamation and the Ethics of Oeconomia*, ed. C. Behrmann (De Gruyter, 2016); Pollmann, *Memory in Early Modern Europe*; Judith Pollmann and Erika Kuijpers, "Introduction: On the Early Modernity of Modern Memory," in *Memory before Modernity: Practices of Memory in Early Modern Europe*, ed. Erika Kuijpers et al. (Brill, 2013).

77. For a broader examination of the long-lasting impact of the trial, both in temporal and geographical terms, see Finocchiaro, *Retrying Galileo*.

78. *GO* 19.535–37, 559–62.

79. On the history of scholarly burial rites, see Marco Beretta, "Heroes, Martyrs and Saints: The Perilous Fate of Savant Relics," in *Brains and Remains of Scientists*, ed. Marco Beretta et al. (Science History Publications, 2016), 1–10.

80. Torricelli, *Lezioni accademiche*, 59: "in cambio d'un Nerone s'immagini un Augusto; per um empio, vizioso, e traditore, un buono, un virtuoso, un fedele."

81. Torricelli, *Lezioni accademiche*, 57: "Se quelli, i quali lo vedono presenzialmente, non lo conoscono, come faranno poi a conoscerlo quelli, che son per nascere di qui a mill'anni?" Torricelli uses the words *infamia* and *infamis* in their broader sense of a social, rather than legal, standing.

82. Jon R. Snyder, *Dissimulation and the Culture of Secrecy in Early Modern Europe* (University of California Press, 2010). Earlier research has shown that such strategies were used by Galileo and his supporters in correspondence; see Hannah Marcus and Paula Findlen, "Deciphering Galileo: Communication and Secrecy Before and After the Trial," *Renaissance Quarterly* 72, no. 3 (2019): 953–95.

83. The most complete account is Findlen, "Long After the Trial."

84. Nicolas-Claude Fabri de Peiresc to Francesco Barberini, 8 December 1634, *GO* 16.170: "ch'ella si degnarà far qualche officio per la consolatione d'un buon vecchio settuagenario et poco sano di corpo, la cui memoria difficilmente sarà scancellata nell'avenire."

85. Nicolas-Claude Fabri de Peiresc to Francesco Barberini, 8 December 1634, GO 16.170: "Sarà difficile che la posterità non gli mostri

sempre grand'obligo delle mirabili noticie da lui scoperte nel cielo con gli suoi occhiali et con l'acutissimo suo ingegno."

86. Translation from Finocchiaro, *Retrying Galileo*, 54. For a more elaborate discussion of these comparisons, see Horst Bredekamp, "Galileo as the Unpunished Artist: Peiresc's Argument," in *Il caso Galileo: Una rilettura storica, filosofica, teologica*, vol. 2, ed. Massimo Bucciantini et al. (Leo S. Olschki, 2011).

87. Nicolas-Claude Fabri de Peiresc to Francesco Barberini, 8 December 1634, *GO* 16.170: "così pare che i secoli a venire potranno trovare stranno, che doppo la ritrattatione d'una opinione che ancora non era stata sssolutamente prohibita in publico nè proposta se non come problematica, si usi tanto rigore ad un povero vecchio settuagenario di tenerlo in carcere, sia pubblico o privato, in maniera che non gli sia lecito di tornare alla città et alla casa sua nè di ricevere le visite et consolationi degli amici, stante le infermità quasi inseparabili della vecchiaia et le necessità delli soccorsi che vi occorono quasi di continuo, che ben spesso non patiscono la dilatione del tempo, che ricchiede la strada et distanza della villa alla città, per i rimedii ad accidenti subitanei."

88. Nicolas-Claude Fabri de Peiresc to Francesco Barberini, 8 December 1634, *GO* 16.170: "Veramente sarà cosa trovata durissima per tutto, et maggiormente dalla posterità che dal secolo presente, dove pare che ogniuno lasci gli interessi del publico, et specialmente delli miseri, per attendere alli proprii. Et sarà appunto una macchia allo splendore et fama di questo Ponteficato."

89. The comparison with Socrates was made by others as well; see, for instance, *GO* app. 2.323.

90. Nicolas-Claude Fabri de Peiresc to Francesco Barberini, 31 January 1635, *GO* 16.202: "et sicuro che sì come l'indulgenza ch'ella farà concedere al suo peccato di fragilità humana sarà conforme alli voti delli più nobili ingegni del secolo, che compatiscono tanto alla severità et prolungatione del suo castigo, così un evento contrario correbbe gran rischio d'essere interpretato e forzi comparato un giorno alla persecutione della persona et sapienza di Socrate nella sua patria, tanto biasimata dall'altre nazioni et dalli posteri istessi di que'che gli diedero tanti travagli." Peiresc's reference to the negative verdict of other nations in particular may be a reference to the reception of Galileo's verdict in France; see Lisa Sarasohn, "French Reaction to the Condemnation of Galileo, 1632–1642," *Catholic Historical Review* 74, no. 1 (1988): 34–54.

91. Sarasohn, "French Reaction to the Condemnation of Galileo," 35; Finocchiaro, *Retrying Galileo*, 53.

92. On Barberini's missing signature, see Finocchiaro, *Retrying Galileo*, 13; and Mayer, *Trying Galileo*, 208.

93. Finocchiaro, *Retrying Galileo*, 55.

94. Muir, *Culture Wars of the Late Renaissance*, 56–57; Jean-Pierre Cavaillé, *Dis/simulations: Jules-César Vanini, François La Mothe Le Vayer, Gabriel Naudé, Louis Machon et Torquato Accetto; Religion, morale et politique au XVII*[e] *siècle* (H. Champion, 2002), 199–266; Snyder, *Dissimulation*, xv.

95. On Dal Pozzo's contributions to science and his relation to Galileo, the Barberini family, and the rest of the Lincei, see Freedberg, *Eye of the Lynx*.

96. On Dal Pozzo's collection, see Simone Testa, *Italian Academies and Their Networks, 1525–1700: From Local to Global* (Palgrave Macmillan, 2015), 130–33.

97. Gabriel Naudé, *Epigrammata in virorum literatorum imagines quas illustrissimus eques Cassianus a Puteo sua in biblioteca dedicavit cum apendicula variorum carminum* (Rome, 1641), A2v: "Non vultum, Galilaee, tuum mihi cura videndi est. / Ast oculata magis picta tabella placet: / Namque oculis reserata tuis qui sidera vidi, / Et Coelo per te reddita iura novo. / Nunc oculos caeca dudum sub nocte latentes, / Aequa non possem cernere mente tuos."

98. Logeion (online Greek and Latin dictionary), "nox," Lewis and Short's *Latin-English Lexicon* (1879), II B, https://logeion.uchicago.edu/nox.

99. Galileo to Dal Pozzo, 20 January 1641, *GO* 18.290: "Non so qual sia maggiore, o il guadagno appresso il mondo della mia reputazione, o lo scapito del purgatissimo giudizio di V. S. Ill.ma, mentre che, da soverchio affetto trasportata, mi colloca in quell'altezza di luogo dove per me già mai non sarei salito." It seems strange that Galileo contacted Dal Pozzo rather than Naudé, yet Naudé's epigrams were written to accompany Dal Pozzo's collection of portraits, and Dal Pozzo and Galileo were already closely acquainted: They were both members of the Accademia dei Lincei.

100. Hannah Marcus, in her discussion of Naudé's library politics, points out that Naudé was open to the idea of having works by heretics available in libraries, while also arguing that access to such books should be limited. Marcus, *Forbidden Knowledge*, 224–25.

101. Naudé's little book was reprinted shortly after it appeared in Rome, for reasons that seem vague. Only two days after Galileo sent his jubilant letter of thanks to Dal Pozzo, the correspondent who had sent him the work wrote again, enclosing a reprint and asking Galileo to return his original copy. The explanation given for the reprint was as follows: "He [Naudé] is not satisfied with the published Epigrams, as, due to the carelessness of the copyist, some mistakes of syllable have

appeared, which he has thus corrected and reprinted, as you will see." Fortunio Liceti to Galileo, 22 January 1641, *GO* 18.291: "Non resta sodisfatto degli Epigrammi publicati, sendovi trascorso, per inavertenza del copista, qualche errore di sillaba, che poi ha corretto e ristampato, come vedrà; e mi sarà favore che V. S. mi rimandi quel foglio, da inviare all'autore che lo ricchiede."

102. Gattei, *On the Life of Galileo*, xxvi–xxix.

103. Findlen, "Long After the Trial," 228.

104. The original Latin reads as follows: "Quod hinc Veritatem et Iustitiam esse fortiter propugnandas Mendacium Assentationem et Hypocrisin veluti pestes defugiendas." For the full Latin transcription and Italian translation, see Roberto Lunardi and Oretta Sabbatini, eds., *Il rimembrar delle passate cose: Una casa per memoria; Galileo e Vincenzo Viviani* (Polistampa, 2009), 29–37. In the rest of the long epitaph, Viviani lists Galileo's achievements, emphasizes his relations with important patrons, and stresses the link between Michelangelo's death and Galileo's birth. Literature on the epitaph includes Galluzzi, "Sepulchers of Galileo"; and Gattei, *On the Life of Galileo*.

105. Vaccalluzzo, *Galileo nella poesia*, 129–31. The poem was included in collections of Maffeo Barberini's poems (*Maphaei S.R.E. card. Barberini nunc Urbani P.P. VIII. Poemata* [Rome, 1631] and *Maphaei S.R.E. card. Barberini nunc Urbani P.P. VIII. Poemata* [Antwerp, 1634]), albeit without reference to Galileo.

106. Giuseppe Gaudenzi, *Paganino Gaudenzi* (Peter Lang, 1975) 3.

107. Paganino Gaudenzi, *In morte del famosissimo Galileo tre sonetti* [n.p., 1642], 44r: "Non pensi alcun, che col lodarsi il Galileo s'approvi la mobilità della terra da lui Accademicamente ne' suoi Dialoghi proposta, o altra oppenione che potesse esser contraria alla determinazione di S. Chiesa."

108. Gaudenzi to Cassiano dal Pozzo, 1 February 1642, *GO* app. 2.468: "La stampa va lentamente ed io per non perder in queste vacanze del Carnovale affatto il tempo ho composto molti sonetti, cinque de' quali invio a V.S. Ill.ma con dirle però che quelli sopra il Galileo, fatti di gusto di S.A.S., non vorrei ch'uscissero dalle mani di V.S. Ill.ma per le cagioni che sa."

109. Jean-Jacques Bouchard, *Nicolai Claudii Fabricii Peirescii senatoris acquensis laudatio habita in funebri concione Academicorum Romanorum* (Venice, 1638), 20.

110. Jean-Jacques Bouchard to Vincenzo Capponi, 14 August 1638, *GO* app. 2.382.

111. Peter Rietbergen, *Power and Religion in Baroque Rome: Barberini Cultural Politics* (Brill, 2006), 412–17. Bouchard's letter indicates that

the order to rewrite the eulogy was given by Niccolò Riccardi, otherwise known as Father Mostro.

112. Bouchard to Vincenzo Capponi, 20 February 1638, *GO* 17.298–99: "Questo le so ben dire, ch'io sono talmente sdegnato di questa barbarie, usata contro il povero Galilei in particolare, ch'io son risoluto d'impiegar il primo tempo libero che mi sarà concesso, a scrivere la sua vita, della quale la prego di voler procurarmi le memorie più particolari che sarà possibile." Bouchard changed his words to "Galilaeus Galileius, rerum superarum atque coelestium, acer & verè lynceus contemplator, Fabricioque nostro amicissimus." Bouchard, *Nicolai Claudii Fabricii Perescii senatoris acquensis laudatio habita in funebri concione Academicorum Romanorum*, in Francesco Barberini, *Monumentum Romanum Nicolao Claudio Fabricio Perescio senatori aquensi doctrinae virtutisque causa factum* (Rome, 1638), 20.

113. Jean-Jacques Bouchard, *Nicolai Claudii Fabricii Peirescii senatoris acquensis laudatio habita in funebri concione Academicorum Romanorum* (Aix-en-Provence, 1639).

114. The original reads as follows: "or vedi, se del Mondo è base e centro / del sol l'immensa sfera, e se la luna / valli, onde, monti tien nel globo dentro." G. Gaudenzi, *Paganino Gaudenzi*, 84.

115. Paganino Gaudenzi, *La Galleria dell'inclito Marino considerata vien dal Paganino, con alcune composizioni dell'istesso Paganino* (Pisa, 1648), 161–65.

116. P. Gaudenzi, *La Galleria dell'inclito Marino*, 162: "Onde se la nobiltà, e grandezza dell'oggetto dà riputazione, e stima a'professori delle scienze, egli sovra ogn'altro in questa età delle celesti contemplazioni si può chiamar benemerito, e si può dir con verità, che finche del Sol, e degli altri Pianeti si favellera, sarà chiara, ed illustre del Galileo la memoria."

117. Rietbergen, *Power and Religion in Baroque Rome*, 274.

118. P. Gaudenzi, *La Galleria dell'inclito Marino*, 165: "Col lodar del Galileo il saper nelle Mattematiche non s'approva la mobilità della terra, della quale egli parlò nei suoi dialoghi."

119. The text has recently been published and translated in Gattei, *On the Life of Galileo*, 1–94.

120. Others have emphasized the similarities between Viviani's work and Vasari's; see especially Michael Segre, "Vite di scienziati, vite di artisti," in *Firenze milleseicentoquaranta: Arti, lettere, musica, scienza*, ed. Elena Fumagalli et al. (Marsilio Editori, 2010); and Gattei, *On the Life of Galileo*, xvii–xxiv.

121. Giorgio Vasari, *The Lives of the Artists*, trans. and ed. Julia Conaway Bondanella and Peter Bondanella (Oxford University Press, 1991), 414–15.

122. Vincenzo Viviani, *Racconto Istorico* in *GO* 19.597–632, 617: "Ma

essendosi già il Sig.[r] Galileo per l'altre sue ammirabili speculazioni con immortal fama sin al cielo inalzato, e con tante novità acquistatosi tra gl'uomini del divino, permesse l'Eterna Providenza ch'ei dimostrasse l'umanità sua con l'errare."

123. This was a fairly common narrative; see Pollmann and Kuypers, "Introduction: On the Early Modernity of Modern Memory," 20.

124. Viviani, *Racconto Istorico*, 617: "onde tornò alla sua villa d'Arcetri, nella quale egli già abitava più del tempo, come situata in buon'aria et assai comoda alla città di Firenze, e perciò facilmente frequentata dalle visite delli amici e domestici, che sempre gli furono di particolar sollievo e consolazione."

125. Michael Segre, "The Never-Ending Galileo Story," in *The Cambridge Companion to Galileo*, ed. Peter Machamer (Cambridge University Press, 2006), 390; Segre, *In the Wake of Galileo*, 107–26; Michael Segre, "Viviani's Life of Galileo," *Isis* 80, no. 2 (1989): 206–31.

126. Another example of this strategy can be found in the work of Leone Allacci, who had already completed the first draft of a book on illustrious men when news of the trial reached him. He then rewrote the section on Galileo, focusing on Galileo's earlier publications and leaving out any reference to the *Dialogo* and trial. Thomas Cerbu and Michel-Pierre Lerner, "La disgrâce de Galilée dans les *Apes Urbanae*: Sur la fabrique du texte de Leone Allacci," *Nuncius* 15, no. 2 (2000): 589–610.

127. Finocchiaro, *Retrying Galileo*, 56.

128. Maurizio Torrini, "'Che il mio nome non si estingua': La morte di Galileo e le sorti della scienza,'" in *Firenze milleseicentoquaranta: Arti, lettere, musica, scienza*, ed. Elena Fumagalli et al. (Marsilio, 2010), 15–36, 21.

129. See Tognoni, *GO* app. 1.49–51. The Palazzo Pitti also includes a fresco, painted by Anton Domenico Gabbiani in 1692–1693, which depicts Galileo surrounded by mathematical instruments and deities associated with wisdom and ingenuity. This fresco, however, was also in the private wings of the Medici palace. Thomas Frangenberg, "A Private Homage to Galileo: Anton Domenico Gabbiani's Frescoes in the Pitti Palace," *Journal of the Warburg and Courtauld Institutes* 59 (1996): 245–73.

130. Tognoni, *GO* app. 1.84–86.

131. Tognoni, *GO* app. 1.50–51.

132. Viviani and Giovan Battista Nelli were instrumental in the eventual construction of the tomb: Viviani left instructions in his final testament, asking to be laid to rest with Galileo in a tomb of his own design, and Nelli carried out his wishes. Viviani was certainly not the only student who erected a posthumous monument for his teacher; see Marisa Bass, "Justus Lipsius and His Silver Pen," *Journal of the Warburg and Courtauld Institutes* 70, no. 1 (2007): 157–94, esp. 186–89, for a beautiful

discussion of the monument Johannes Wolverius posthumously erected for Justus Lipsius in the Basilica of Our Lady in Halle.

133. For a discussion of the iconography of the tomb, see Arjan de Koomen, "Manipulating Memories: Postponed Tombs for Galileo and Machiavelli," in *Memory and Oblivion*, vol. 1, ed. Wessel Reinink and Jeroen Stumpel (Springer, 1999); and Galluzzi, "I sepolcri di Galileo."

134. By 1674, Viviani had already placed a plaque commemorating Galileo's achievements at length near the Galilei family grave. Galluzzi, "I sepolcri di Galileo," 156; Galluzzi, "Sepulchers of Galileo," 426.

135. The Latin inscription and Italian translation are in Lunardi and Sabbatini, *Il rimembrar delle passate cose*, 256–59.

136. Gattei, "From Banned Mortal Remains to the Worshipped Relics of a Martyr of Science," 79.

137. Findlen, "Long After the Trial," 231–32.

138. Stefano Gattei offers a lively and detailed account of the reburial and discusses the fate of the different relics; see Gattei, "From Banned Mortal Remains to the Worshipped Relics of a Martyr of Science," 79–84.

139. Segre, *In the Wake of Galileo*; Massimo Bucciantini, ed., *The Science and Myth of Galileo Between the Seventeenth and Nineteenth Centuries in Europe* (Leo S. Olschki, 2021).

EPILOGUE

1. Grafton, "Sketch Map of a Lost Continent"; Steven Shapin, *The Scientific Revolution* (University of Chicago Press, 1996).

2. J. B. Shank, "After the Scientific Revolution: Thinking Globally About the Histories of the Modern Sciences," *Journal of Early Modern History* 21, no. 5 (October 2017): 387–88.

3. Antonio Godoli, Francesco Palla, and Alberto Righini, *La villa di Galileo in Arcetri / Galileo's Villa at Arcetri* (Firenze University Press, 2016), 77.

4. Rebekah Higgitt, "Challenging Tropes: Genius, Heroic Invention, and the Longitude Problem in the Museum," *Isis* 108, no. 2 (June 2017): 371–80.

5. "How Can Scientists Make the Most of the Public's Trust in Them?," *Nature* 626, 1 February 2024.

6. For a good discussion of this phenomenon, see the article by science historian Rebekah Higgitt, "Flat-Earthers Aren't the Only Ones Getting Things Wrong," *The Guardian*, 16 January 2016, https://www.theguardian.com/science/the-h-word/2016/jan/21/flat-earthers-myths-science-religion-galileo.

Bibliography

Archives

Diario del Ripieno (1640–1663), Fascetta 76. Archivio storico dell'Accademia della Crusca, Florence.

Fondo Galileiano. Biblioteca Nazionale Centrale di Firenze, Florence.

Mediceo del Principato. Archivio di Stato di Firenze, Florence.

Printed Sources

Accademico Incognito. *Considerazioni sopra il discorso del Sig. Galileo Galilei Intorno alle cose che stanno in su l'acqua o che in quella si muovono, Dedicate alla serenissima D. Maria Maddalena, Archiducessa d'Austria, Gran Duchessa di Toscana: Fatte a difesa, e dichiarazione, dell'opinione d'Aristotile da Accademico Incognito.* Florence, 1612.

Achbari, Azadeh. "The Reviews of *Leviathan and the Air-Pump*: A Survey." *Isis* 108, no. 1 (2017): 108–16.

Assmann, Jan. *Cultural Memory and Early Civilization: Writing, Remembrance, and Political Imagination.* Cambridge University Press, 2012.

Bacon, Francis. *The Essays or Counsels, Civil and Moral.* Edited and introduced by Brian Vickers. Folio Society, 2002.

Baldini, Ugo, and Leen Spruit, eds. *Catholic Church and Modern Science: Documents from the Archives of the Roman Congregations of the Holy Office and the Index.* Vol. 1, *Sixteenth-Century Documents*, book 1. Libreria Editrice Vaticana, 2009.

Barnes, Susan. "The Uomini Illustri, Humanist Culture, and the Development of a Portrait Tradition in Early Seventeenth-Century Italy." *Studies in the History of Art* 27 (1989): 80–92.

Barsanti, Danilo. "I docenti e le cattedra dal 1543 al 1737." In *Storia dell'università di Pisa, 1343–1737*, edited by Commissione rettorale dell'Università di Pisa. Vol. 1, pt. 2. Pacini Editore, 2000.

Bass, Marisa. "Justus Lipsius and His Silver Pen." *Journal of the Warburg and Courtauld Institutes* 70, no. 1 (2007): 157–94.

Battistini, Andrea. "'Cedat Columbus' e 'Vicisti, Galilaee!': Due esploratori a confronto nell'immaginario barocco." *Annali d'Italianistica* 10 (1992): 116–32.

Battistini, Andrea. *Galileo e i Gesuiti: Miti letterari e retorica della scienza.* Vita e Pensiero, 2000.

Beller, Manfred, and Joep Leerssen, eds. *Imagology: The Cultural Construction and Literary Representation of National Characters: A Critical Survey.* Brill, 2007.

Beretta, Francesco. "Galilée devant le tribunal de l'Inquisition: Une relecture des sources." Université de Fribourg [dissertation], 1998.

Beretta, Francesco. "Les dominicains et le procès de Galilée, ou de l'Inquisition comme instrument de promotion sociale et d'hégémonie intellectuelle." In *I Domenicani e l'Inquisizione Romana: Atti del 3. Seminario internazionale di studi su i domenicani e l'Inquisizione; Roma, 15–18 febbraio*, edited by Carlo Longo. Istituto Storico Domenicano, 2008.

Beretta, Marco. "Heroes, Martyrs and Saints: The Perilous Fate of Savant Relics." In *Brains and Remains of Scientists*, edited by Marco Beretta, Maria Conforti, and Paolo Mazzarello. Science History Publications, 2016.

Biagioli, Mario. *Galileo Courtier: The Practice of Science in the Culture of Absolutism.* University of Chicago Press, 1994.

Biagioli, Mario. *Galileo's Instruments of Credit: Telescopes, Images, Secrecy.* University of Chicago Press, 2006.

Biagioli, Mario. "Galileo the Emblem-Maker." *Isis* 81, no. 2 (1990): 230–58.

Blackwell, Richard. *Behind the Scenes at Galileo's Trial.* University of Notre Dame Press, 2008.

Blackwell, Richard. *Galileo, Bellarmine, and the Bible: Including a Translation of Foscarini's Letter on the Motion of the Earth.* University of Notre Dame Press, 1992.

Blair, Ann. "Natural Philosophy." In *The Cambridge History of Science.* Vol. 3, *Early Modern Science*, edited by Katherine Park and Lorraine Daston. Cambridge University Press, 2006.

Blair, Ann. "Tradition and Innovation in Early Modern Natural Philosophy: Jean Bodin and Jean-Cécile Frey." *Perspectives on Science* 2, no. 4 (1994): 428–54.

Bond, Sarah. "Altering Infamy: Status, Violence, and Civic Exclusion in Late Antiquity." *Classical Antiquity* 33, no. 1 (2014): 1–30.

Botelho, Keith M. *Renaissance Earwitnesses: Rumor and Early Modern Masculinity.* Palgrave Macmillan, 2009.

Bots, Hans, and Françoise Waquet. *La République des Lettres*. De Boeck, 1997.

Bouchard, Jean-Jacques. *Nicolai Claudii Fabricii Peirescii senatoris acquensis laudatio habita in funebri concione Academicorum Romanorum*. Venice, 1638.

Bouchard, Jean-Jacques. *Nicolai Claudii Fabricii Peirescii senatoris acquensis laudatio habita in funebri concione Academicorum Romanorum*. Aix-en-Provence, 1639.

Bouchard, Jean-Jacques. *Nicolai Claudii Fabricii Perescii senatoris acquensis laudatio habita in funebri concione Academicorum Romanorum*. In *Monumentum Romanum Nicolao Claudio Fabricio Perescio senatori aquensi doctrinae virtutisque causa factum*, compiled by Francesco Barberini. Rome, 1638.

Bowman, Jeffrey A. "Infamy and Proof in Medieval Spain." In *Fama: The Politics of Talk and Reputation in Medieval Europe*, edited by Thelma Fenster and Daniel Lord Smail. Cornell University Press, 2003.

Braudy, Leo. *The Frenzy of Renown: Fame and Its History*. Oxford University Press, 1986.

Bredekamp, Horst. "Galileo as the Unpunished Artist: Peiresc's Argument." In *Il caso Galileo: Una rilettura storica, filosofica, teologica*, vol. 2, edited by Massimo Bucciantini, Michele Camerota, and Franco Giudice. Leo S. Olschki, 2011.

Brémond, Claude, Jacques Le Goff, and Jean-Claude Schmitt. *"L'exemplum."* Brepols, 1982.

Brevaglieri, Sabina. "Science, Books and Censorship in the Academy of the Lincei." In *Conflicting Duties: Science, Medicine and Religion in Rome (1550–1750)*, edited by Maria Pia Donato and Jill Kraye. University of London Press, 2009.

Brown, Alison. "Defining the Place of Academies in Florentine Culture and Politics." In *The Italian Academies 1525–1700: Networks of Culture, Innovation and Dissent*, edited by Jane E. Everson, Denis V. Reidy, and Lisa Sampson. Routledge, 2016.

Browne, Janet E. "Charles Darwin as a Celebrity." *Science in Context* 16, no. 1–2 (2003): 175–94.

Browne, Janet E. "Looking at Darwin: Portraits and the Making of an Icon." *Isis* 100, no. 3 (2009): 542–70.

Bucciantini, Massimo. *Contro Galileo: Alle origini dell'affaire*. Leo S. Olschki, 1995.

Bucciantini, Massimo, ed. *The Science and Myth of Galileo Between the Seventeenth and Nineteenth Centuries in Europe*. Leo S. Olschki, 2021.

Bucciantini, Massimo, Michele Camerota, and Franco Giudice. *Galileo's*

Telescope: A European Story. Translated by Catherine Bolton. Harvard University Press, 2015.

Burke, Peter. "Erasmus and the Republic of Letters." *European Review* 7, no. 1 (1999): 5–17.

Burke, Peter. *The Fabrication of Louis XIV*. Yale University Press, 1992.

Burke, Peter. "Renaissance Individualism and the Portrait." *History of European Ideas* 21, no. 3 (1995): 393–400.

Calendar of State Papers Relating to English Affairs in the Archives of Venice. Vol. 20, *1626–1628*, edited by Allen B. Hinds. His Majesty's Stationery Office, 1914.

Camerota, Michele. *Galileo Galilei e la cultura scientifica nell'età della Controriforma*. Salerno Editrice, 2004.

Camerota, Michele. "Per segno e per uffizio di animo pronto e leale: Il trattato antigalileiano dell'Accademico Incognito." In *Annali della Facoltà di scienze della formazione dell'Università di Cagliari*. Vol. 23. Cagliari, 2000.

Camerota, Michele, Franco Giudice, and Salvatore Ricciardo. "The Reappearance of Galileo's Original Letter to Castelli." *Notes and Records: The Royal Society Journal for the History of Science* 73, no. 1 (March 2019): 11–28.

Capra, Baldassare. *Usus et fabrica circini cuiusdam proportionis, opera et studio*. Padua, 1607.

Carabba, Carlo, and Giuliano Gasparri. "La vita di Girolamo Magagnati." In Girolamo Magagnati, *Lettere a diversi*, edited and introduced by Laura Salvetti Firpo. Leo S. Olschki, 2006.

Carter, Sarah. *Ovidian Myth and Sexual Deviance in Early Modern English Literature*. Springer, 2011.

Cavaillé, Jean-Pierre. *Dis/simulations: Jules-César Vanini, François La Mothe Le Vayer, Gabriel Naudé, Louis Machon et Torquato Accetto; Religion, morale et politique au XVIIe siècle*. H. Champion, 2002.

Cerbu, Thomas, and Michel-Pierre Lerner. "La disgrâce de Galilée dans les *Apes Urbanae*: Sur la fabrique du texte de Leone Allacci." *Nuncius* 15, no. 2 (2000): 589–610.

Chang, Ku-ming (Kevin). "From Oral Disputation to Written Text: The Transformation of the Dissertation in Early Modern Europe." In *History of Universities*, vol. 19, edited by Mordechai Feingold. Oxford University Press, 2004.

Clark, William. *Academic Charisma and the Origins of the Research University*. University of Chicago Press, 2006.

Cohen, H. Floris. *How Modern Science Came into the World: Four Civilizations, One 17th-Century Breakthrough*. Amsterdam University Press, 2010.

Cole, Janie. "Cultural Clientelism and Brokerage Networks in Early Modern Florence and Rome: New Correspondence Between the Barberini and Michelangelo Buonarroti the Younger." *Renaissance Quarterly* 60, no. 3 (2007): 729–88.

Coresio, Giorgio. *Operetta intorno al galleggiare de corpi solidi*. Florence, 1612.

Costa, Emilio. "Galileo e lo Studio di Bologna." *Studi e Memorie per la Storia dell'Università di Bologna* 7 (1992): 8–11.

Cozzi, Gaetano. *Paolo Sarpi tra Venezia e l'Europa*. Einaudi, 1979.

Crowston, Clare Haru. *Credit, Fashion, Sex: Economies of Regard in Old Regime France*. Duke University Press, 2013.

Dall'Aglio, Stefano. "'Faithful to the Spoken Word': Sermons from Orality to Writing in Early Modern Italy." *The Italianist* 34, no. 3 (2014): 463–77.

Dall'Aglio, Stefano. "Voices Under Trial: Inquisition, Abjuration, and Preachers' Orality in Sixteenth-Century Italy." *Renaissance Studies* 31, no. 1 (2017): 25–42.

Darnton, Robert. *The Forbidden Best-Sellers of Pre-Revolutionary France*. Norton, 1996.

Daston, Lorraine. "The Ideal and Reality of the Republic of Letters in the Enlightenment." *Science in Context* 4, no. 2 (1991): 367–86.

Daston, Lorraine. *Rivals: How Scientists Learned to Cooperate*. Columbia Global Reports, 2023.

De Angelis, Alessandro. "Galileo Galilei and a Forgotten Poem on the 1604 Supernova." *Quaderni di Storia della Fisica* 1, no. 1 (2023): 17–29.

De Ceglia, Francesco. *De Natantibus: Una disputa ai confini tra filosofia e matematica nella Toscana medicea (1611–1615)*. Edizioni Giuseppe Laterza, 1999.

De Ceglia, Francesco. "Giorgio Coresio: Note in merito a un difensore dell'opinione di Aristotele." *Physis* 37, no. 2 (2000): 293–437.

De Koomen, Arjen. "Manipulating Memories: Postponed Tombs for Galileo and Machiavelli." In *Memory and Oblivion*, vol. 1, edited by Wessel Reinink and Jeroen Stumpel. Springer, 1999.

Delle Colombe, Ludovico. *Discorso Apologetico di Lodovico delle Colombe, d'Intorno al Discorso di Galileo Galilei*. Florence, 1612.

Delle Colombe, Raffaele. *Delle Prediche sopra tutti gli Evangeli dell'anno, nelle quali con similitudini, metafore, allegorie retoriche, e considerazioni particolari si dichiarano molti luoghi morali della Sacra Scrittura, ma in senso letterale: Di fra Raffaello delle Colombe dell'Ordine de' predicatori; Con tre tauole . . . Volume primo*. Florence, 1613.

Delle Colombe, Raffaele. *Dupplicato avvento di prediche. Il primo à perso-*

ne religiose. Il secondo comunemente a tutti. Del padre fra Raffaello delle Colombe domenicano. Con tre tavole copiose . . . Quarto tomo delle sue prediche. Florence, 1627.

Delle Colombe, Raffaele. *Prediche della Quaresima con esposizioni di Scritture sacre, Varieta di traslazioni, Dottrine morali, Diuerse erudizioni, e Similitudini, di fra Raffaello Delle Colombe Domenicano, Con Tauole copiose; Delle Scritture esposte, delle cose notabili, e De' titoli di ciascuna Predica . . . 2: Delle prediche di tutto l'anno*. Florence, 1615 and 1622.

De Martino, Domenico, ed. *Lezioni accademiche d'Evangelista Torricelli*. Biblion, 2009.

Dietz Moss, Jean. "Galileo's Letter to Christina: Some Rhetorical Considerations." *Renaissance Quarterly* 36, no. 4 (Winter 1983): 547–76.

Dietz Moss, Jean. *Novelties in the Heavens: Rhetoric and Science in the Copernican Controversy*. University of Chicago Press. 1993.

Di Grazia, Vincenzo. *Considerazioni di M. Vincenzo di Grazia sopra'l Discorso di Galileo Galilei intorno alle cose che stanno in su l'acqua, e che in quella si muovono*. Florence, 1613.

Donahue, William. "Astronomy." In *The Cambridge History of Science*. Vol. 3, *Early Modern Science*, edited by Katherine Park and Lorraine Daston. Cambridge University Press, 2006.

Drake, Stillman. *Cause, Experiment, and Science: A Galilean Dialogue Incorporating a New English Translation of Galileo's "Bodies That Stay atop Water, or Move in It."* University of Chicago Press, 1981.

Drake, Stillman. *Galileo at Work: His Scientific Biography*. University of Chicago Press, 1978.

Drake, Stillman. "Galileo Gleanings IV: Bibliographical Notes." *Isis* 49, no. 4 (December 1958): 409–13.

Drake, Stillman. "Galileo Gleanings XIV: Galileo and Girolamo Magagnati." *Physis* 6, no. 3 (1964): 269–86.

Drake, Stillman. *Galileo Studies: Personality, Tradition, and Revolution*. University of Michigan Press, 1970.

Egmond, Florike. "Execution, Dissection, Pain, and Infamy—A Morphological Investigation." In *Bodily Extremities: Preoccupations with the Human Body in Early Modern European Culture*, edited by Florike Egmond and Robert Zwijnenberg. Routledge, 2003.

Enenkel, Karl. *Die Stiftung von Autorschaft in der neulateinischen Literatur (ca. 1350–ca. 1650): Zur autorisierenden und wissensvermittelnden Funktion von Widmungen, Vorworttexten, Autorporträts und Dedikationsbildern*. Brill, 2015.

Enenkel, Karl A. E., and Konrad Adriaan Ottenheym, eds. *The Quest for an Appropriate Past in Literature, Art and Architecture*. Brill, 2018.

Fara, Patricia. "Framing the Evidence: Scientific Biography and Portraiture." In *The History and Poetics of Scientific Biography*, edited by Thomas Söderqvist. Routledge, 2007.

Fara, Patricia. *Newton: The Making of Genius*. Columbia University Press, 2002.

Favaro, Antonio. *Amici e corrispondenti di Galileo*. Edited by Paolo Galluzzi. Libreria editrice Salimbeni, 1983.

Favaro, Antonio. "Galileo ed il Magini: Aspiranti ad una lettura di matematica nello Studio di Bologna." *Atti del R. Istituto veneto di scienze, lettere ed arti* 82 (1922–1923): 145–55.

Feingold, Mordechai. "*Fama*: Les savants jésuites et la quête de la renommée." *Dix-Septième Siècle* 237, no. 4 (2007): 755–74.

Feingold, Mordechai. "The Grounds for Conflict: Grienberger, Grassi, Galileo, and Posterity." In *The New Science and Jesuit Science: Seventeenth Century Perspectives*, edited by Mordechai Feingold. Springer, 2003.

Feldhay, Rivka. *Galileo and the Church: Political Inquisition or Critical Dialogue?* Cambridge University Press, 1995.

Fenster, Thelma, and Daniel Lord Smail, eds. *Fama: The Politics of Talk and Reputation in Medieval Europe*. Cornell University Press, 2003.

Findlen, Paula. "Long After the Trial: Galileo's Rediscovery, Florentine Nostalgia, and Enlightened Passions." In *Florence After the Medici: Tuscan Enlightenment, 1737–1790*, edited by Corey Tazzara, Paula Findlen, and Jacob Soll. Routledge, 2019.

Findlen, Paula. "Rethinking 1633: Writing the Life of Galileo after the Trial." In *Nature Engaged: Science in Practice from the Renaissance to the Present*, edited by Mario Biagioli and Jessica Riskin. Palgrave, 2012.

Fine, Gary Alan. *Difficult Reputations: Collective Memories of the Evil, Inept, and Controversial*. University of Chicago Press, 2001.

Finocchiaro, Maurice A. *The Galileo Affair: A Documentary History*. University of California Press, 1989.

Finocchiaro, Maurice A. *Retrying Galileo, 1633–1992*. University of California Press, 2005.

Flower, Harriet I. *The Art of Forgetting: Disgrace and Oblivion in Roman Political Culture*. University of North Carolina Press, 2006.

Fontaine, Laurence. *The Moral Economy: Poverty, Credit, and Trust in Early Modern Europe*. Cambridge University Press, 2014.

Fosi, Irene. *Papal Justice: Subjects and Courts in the Papal States, 1500–1750*. Translated by Thomas V. Cohen. Catholic University of America Press, 2007.

Foucault, Michel. *Discipline and Punish: The Birth of the Prison*. Pantheon Books, 1977.

Frangenberg, Thomas. "A Private Homage to Galileo: Anton Domenico

Gabbiani's Frescoes in the Pitti Palace." *Journal of the Warburg and Courtauld Institutes* 59 (1996): 245–73.

Freedberg, David. *The Eye of the Lynx: Galileo, His Friends, and the Beginnings of Modern Natural History*. University of Chicago Press, 2002.

Frisch, Andrea. *The Invention of the Eyewitness: Witnessing and Testimony in Early Modern France*. University of North Carolina Press, 2004.

Füssel, Marian. "Die Gelehrtenrepublik im Kriegszustand: Zur bellizitären Metaphorik von gelehrten Streitkulturen der Frühen Neuzeit." In *Gelehrte Polemik: Intellektuelle Konfliktverschäftungen um 1700*, edited by Kai Bremer and Carlos Spoerhase. Vittorio Klostermann, 2011.

Füssel, Marian. "On the Means of Becoming Famous in the Learned World." In *Scholars in Action: The Practice of Knowledge and the Figure of the Savant in the 18th Century*, vol. 1, edited by André Holenstein, Hubert Steinke, and Martin Stuber. Brill, 2013.

Gabrieli, Giuseppe, ed. "Il Carteggio Linceo della vecchia Accademia di Federico Cesi (1603-1630)." *Memorie della R. Accademia dei Lincei*, series 6, Classe di scienze morali, storiche e filologiche, 7 (1938–1942), part II (1610–1624), 350–51.

Gage, Frances. "Caravaggio's *Rumore*: Fact, Fiction and Authority in Giovanni Baglione's *Lives of the Painters, Sculptors and Architects*." *Past & Present* 257, supplement 16 (2022): 111–40.

Galilei, Galileo. *Difesa di Galileo Galilei contro alle calunnie & imposture di Baldassar Capra Milanese*. Venice, 1607.

Galilei, Galileo. *Discorso al Serenissimo Don Cosimo II Gran Duca di Toscana intorno alle cose che stanno in sull'acqua o in quello si muovono, di Galileo Galilei Filosofo, e Matematico della Medesima Altezza Serenissima*. Florence, 1612.

Galilei, Galileo. *Istoria e dimostrazioni intorno alle macchie solari e loro accidenti comprese in tre lettere scritte a Marco Velseri*. Rome, 1613.

Galilei, Galileo. *Le operazioni del compasso geometrico e militare*. Padua, 1606.

Galilei, Galileo. *Le opere di Galileo Galilei*. Edited by Antonio Favaro. 20 vols. Florence, Barbèra, 1890–1909; reprint, 1929–1939.

Galilei, Galileo. *Le opere di Galileo Galilei: Appendice*. Vol. 1, *Iconografia Galileiana*, edited by Federico Tognoni. Giunti Editore, 2013.

Galilei, Galileo. *Le opere di Galileo Galilei: Appendice*. Vol. 2, *Carteggio*, edited by Michele Camerota and Patrizia Ruffo, in collaboration with Massimo Bucciantini. Giunti Editore, 2015.

Galilei, Galileo. *Sidereus Nuncius; or, the Sidereal Messenger*. 1610. Translated by Albert Van Helden. University of Chicago Press, 1989.

Galilei, Galileo, and Christoph Scheiner. *On Sunspots*. Translated and introduced by Eileen Reeves and Albert Van Helden. University of Chicago Press, 2010.

Galluzzi, Paolo. "I sepolcri di Galileo: Le spoglie 'vive' di un eroe della scienza." In *Il Pantheon di Santa Croce a Firenze*, edited by L. Berti. Cassa di Risparmio, 1993.

Galluzzi, Paolo. *The Lynx and the Telescope: The Parallel Worlds of Federico Cesi and Galileo*. Translated by Peter Mason. Brill, 2017.

Galluzzi, Paolo. "The Sepulchers of Galileo: The 'Living' Remains of a Hero of Science." In *The Cambridge Companion to Galileo*, edited by Peter Machamer. Cambridge University Press, 2006.

Garber, Daniel. "Descartes among the Novatores." *Res Philosophica* 92, no. 1 (2015): 1–19.

Garber, Daniel. "Historicizing Novelty." In *What Reason Promises: Essays on Reason, Nature and History*, edited by Wendy Doniger, Peter Galison, and Susan Neiman. De Gruyter, 2016.

Garber, Daniel. "Telesio among the Novatores: Telesio's Reception in the Seventeenth Century." In *Early Modern Philosophers and the Renaissance Legacy*, edited by Cecilia Muratori and Gianni Paganini, 119–33. Springer, 2016.

Gattei, Stefano. "From Banned Mortal Remains to the Worshipped Relics of a Martyr of Science: Vincenzo Viviani and the Birth of Galileo's Mythography." In *Brains and Remains of Scientists*, edited by Marco Beretta, Maria Conforti, and Paolo Mazzarello. Science History Publications, 2016.

Gattei, Stefano, ed. and trans. *On the Life of Galileo: Viviani's Historical Account and Other Early Biographies*. Princeton University Press, 2019.

Gaudenzi, Giuseppe. *Paganino Gaudenzi*. Peter Lang, 1975.

Gaudenzi, Paganino. *In morte del famosissimo Galileo tre sonetti*. N.p., 1642.

Gaudenzi, Paganino. *La Galleria dell'inclito Marino considerata vien dal Paganino, con alcune composizioni dell'istesso Paganino*. Pisa, 1648.

Gaylard, Susan. *Hollow Men: Writing, Objects, and Public Image in Renaissance Italy*. Fordham University Press, 2003.

Gilmour, David. *The Pursuit of Italy: A History of a Land, Its Regions, and Their Peoples*. Penguin Books, 2011.

Godel, Rainer. "Controversy as the Impetus for Enlightened Practice of Knowledge." In *Scholars in Action: The Practice of Knowledge and the Figure of the Savant in the 18th Century*, vol. 1, edited by André Holenstein, Hubert Steinke, and Martin Stuber. Brill, 2013.

Godoli, Antonio, Francesco Palla, and Alberto Righini. *La villa di Galileo in Arcetri / Galileo's Villa at Arcetri*. Firenze University Press, 2016.

Goldgar, Anne. *Impolite Learning: Conduct and Community in the Republic of Letters, 1680–1750.* Yale University Press, 1995.

Goudriaan, Elisa. *Florentine Patricians and Their Networks: Structures Behind the Cultural Success and the Political Representation of the Medici Court (1600–1660).* Brill, 2018.

Grafton, Anthony. "A Sketch Map of a Lost Continent: The Republic of Letters." *Republics of Letters: A Journal for the Study of Knowledge, Politics, and the Arts* 1 (1 May 2009): 1–18.

Greenblatt, Stephen. *Renaissance Self-Fashioning: From More to Shakespeare.* University of Chicago Press, 1980.

Greenidge, A. H. J. *Infamia: Its Place in Roman Public and Private Law.* Oxford, 1894.

Grendler, Paul F. *The Roman Inquisition and the Venetian Press, 1540–1605.* Princeton University Press, 1977.

Grendler, Paul F. *The Universities of the Italian Renaissance.* Johns Hopkins University Press, 2002.

Guastella, Gianni. *Word of Mouth: Fama and Its Personifications in Art and Literature from Ancient Rome to the Middle Ages.* Oxford University Press, 2017.

Guerrini, Luigi. "The Archbishop and Astronomy: Alessandro Marzimedici and the 1604 Supernova." In *Copernicus Banned: The Entangled Matter of the Anti-Copernican Decree of 1616*, edited by Natacha Fabri and Federica Favino. Leo S. Olschki, 2018.

Guerrini, Luigi. *Cosmologie in lotta: Le origini del processo di Galileo.* Polistampa, 2010.

Guerrini, Luigi. "Echoes from the Pulpit: A Preacher Against Galileo's Astronomy." *Journal for the History of Astronomy* 43, no. 4 (2012): 377–90.

Guerrini, Luigi. *Galileo e la polemica anticopernicana a Firenze.* Polistampa, 2009.

Guerrini, Luigi. "Galileo e Raffaelo delle Colombe." In *Il caso Galileo: Una rilettura storica, filosofica, teologica; Convegno internazionale di studi, Firenze, 26–30 maggio 2009*, vol. 2, edited by Massimo Bucciantini, Michele Camerota, and Franco Giudice. Leo S. Olschki, 2011.

Guerrini, Luigi. "Le *Stanze sopra le stelle e macchie solari scoperte col uovo Occhiale* di Vincenzo Figliucci: Un episodio poco noto della visita di Galileo Galilei a Roma nel 1611." *Lettere Italiane* 50, no. 3 (1998): 387–415.

Guerrini, Luigi. "Raffaello delle Colombe et les origines de la polémique anti-galiléenne à Florence, 1610–1615." In *Il processo a Galileo Galilei e la questione galileiana*, edited by Gian Mario Bravo and Vincenzo Ferrone. Edizioni di storia e letteratura, 2010.

Hacke, Martina. "The Messengers of the Nations of the University of Paris and the Book Trade (Late Fifteenth and Sixteenth Centuries)." In *Early Modern Universities: Networks of Higher Learning*, edited by Anja-Silvia Goeing, Glyn Parry, and Mordechai Feingold. Brill, 2020.

Hall, Crystal. "Galileo, Poetry, and Patronage: Giulio Strozzi's *Venetia edificata* and the Place of Galileo in Seventeenth-Century Italian Poetry." *Renaissance Quarterly* 66, no. 4 (2013): 1296–1331.

Hall, Crystal. *Galileo's Reading*. Cambridge University Press, 2013.

Hardie, Philip. *Rumour and Renown: Representations of Fama in Western Literature*. Cambridge University Press, 2012.

Harkness, Deborah. *The Jewel House: Elizabethan London and the Scientific Revolution*. Yale University Press, 2008.

Härter, Karl. "Images of Dishonoured Rebels and Infamous Revolts: Political Crime, Shaming Punishments and Defamation in the Early Modern Pictorial Media." In *Images of Shame: Infamy, Defamation and the Ethics of Oeconomia*, edited by C. Behrmann. De Gruyter, 2016.

Hedrick, Charles W. *History and Silence: Purge and Rehabilitation in Late Antiquity*. University of Texas Press, 2000.

Heilbron, John. *Galileo*. Oxford University Press, 2010.

Hemmungs Wirtén, Eva. *Making Marie Curie: Intellectual Property and Celebrity Culture in an Age of Information*. University of Chicago Press, 2015.

Hendrix, Harald, ed. *Writers' Houses and the Making of Memory*. Routledge, 2008.

Higgitt, Rebekah. "Challenging Tropes: Genius, Heroic Invention, and the Longitude Problem in the Museum." *Isis* 108, no. 2 (June 2017): 371–80.

Higgitt, Rebekah. *Recreating Newton: Biographies of Newton and the Making of Nineteenth-Century History of Science*. Pickering & Chatto, 2007.

Horky, Martin. *Brevissima peregrinatione contra Nuncium sidereum*. Modena, 1610.

Horodowich, Elisabeth. "The Gossiping Tongue: Oral Networks, Public Life and Political Culture in Early Modern Venice." *Renaissance Studies* 19, no. 1 (2005): 22–45.

Horodowich, Elisabeth. "The Meanings of Gossip in Sixteenth Century Venice." In *Spoken Word and Social Practice: Orality in Europe (1400–1700)*, edited by Thomas V. Cohen and Lesley K. Twomey. Brill, 2015.

Howard, Nicole. *Loath to Print: The Reluctant Scientific Author, 1500–1750*. Johns Hopkins University Press, 2022.

Hunt, Arnold. *The Art of Hearing: English Preachers and Their Audiences, 1590–1640.* Cambridge University Press, 2010.

Inglis, Fred. *A Short History of Celebrity from Byron to Beckham.* Princeton University Press, 2010.

Johns, Adrian. *The Nature of the Book: Print and Knowledge in the Making.* University of Chicago Press, 1998.

Johnson, James H. *Venice Incognito: Masks in the Serene Republic.* University of California Press, 2011.

Kavey, Allison B. "Mercury Falling: Gender Flexibility and Eroticism in Popular Alchemy." In *The Sciences of Homosexuality in Early Modern Europe*, edited by Kenneth Boris and George S. Rousseau. Routledge, 2008.

Keblusek, Marika, and Badeloch Vera Noldus, eds. *Double Agents: Cultural and Political Brokerage in Early Modern Europe.* Brill, 2011.

Kennedy, George A. *Classical Rhetoric and Its Christian and Secular Tradition from Ancient to Modern Times.* 2nd and enlarged ed. University of North Carolina Press, 1999.

Kepler, Johannes. *Kepler's Conversation with Galileo's Sidereal Messenger.* Translated and introduced by Edward Rosen. Johnson Reprint, 1965.

Kivistö, Sari. *The Vices of Learning: Morality and Knowledge at Early Modern Universities.* Brill, 2014.

Kuehn, Thomas. "*Fama* as a Legal Status in Renaissance Florence." In *Fama: The Politics of Talk and Reputation in Medieval Europe*, edited by Thelma Fenster and Daniel Lord Smail. Cornell University Press, 2003.

Landau, Peter. *Die Entstehung des Kanonischen Infamiebegriffs von Gratian bis zur Glossa Ordinaria.* Bohlau, 1966.

Leemans, Inger, and Anne Goldgar, eds. *Early Modern Knowledge Societies as Affective Economies.* Routledge, 2020.

Leerssen, Joep. *National Thought in Europe: A Cultural History.* Amsterdam University Press, 2006.

Lilti, Antoine. *The Invention of Celebrity: 1750–1850.* Translated by Lynn Jeffress. Polity, 2017.

Lines, David. *The Dynamics of Learning in Early Modern Italy: Arts and Medicine at the University of Bologna.* Harvard University Press, 2023.

Lines, David. "Papal Power and University Control in Early Modern Italy: Bologna and Gregory XIII." *Sixteenth Century Journal* 43, no. 3 (2013): 663–82.

Livingstone, David. *Putting Science in Its Place: Geographies of Scientific Knowledge.* University of Chicago Press, 2003.

Lucretius. *On the Nature of Things*. Translated by W. H. D. Rouse, revised by Martin F. Smith. Harvard University Press, 1924.

Lunardi, Roberto, and Oretta Sabbatini, eds. *Il rimembrar delle passate cose: Una casa per memoria; Galileo e Vincenzo Viviani*. Polistampa, 2009.

Lüthy, Christoph. "Atomism, Lynceus, and the Fate of Seventeenth-Century Microscopy." *Early Science and Medicine* 1, no. 1 (1996): 1–27.

Lux, David, and Harold Cook. "Closed Circles or Open Networks? Communicating at a Distance during the Scientific Revolution." *History of Science* 36, no. 2 (1998): 179–211.

Mack, Peter. *A History of Renaissance Rhetoric, 1380–1620*. Oxford University Press, 2011.

Magagnati, Girolamo. *Meditazione poetica sopra i pianeti medicei*. Venice, 1610.

Malagola, Carlo. "Galileo Galilei e L'università di Bologna: Memoria del Dott. Carlo Malagola." *Archivio Storico Italiano*, 4th ser., 7, no. 122 (1881): 187–203.

Maphaei S.R.E. card. Barberini nunc Urbani P.P. VIII. Poemata. Antwerp, 1634.

Maphaei S.R.E. card. Barberini nunc Urbani P.P. VIII. Poemata. Rome, 1631.

Marcus, Hannah. *Forbidden Knowledge: Medicine, Science, and Censorship in Early Modern Italy*. University of Chicago Press, 2020.

Marcus, Hannah, and Paula Findlen. "Deciphering Galileo: Communication and Secrecy Before and After the Trial." *Renaissance Quarterly* 72, no. 3 (2019): 953–95.

Margócsy, Dániel. *Commercial Visions: Science, Trade, and Visual Culture in the Dutch Golden Age*. University of Chicago Press, 2014.

Marr, Alexander. *Between Raphael and Galileo: Mutio Oddi and the Mathematical Culture of Late Renaissance Italy*. University of Chicago Press, 2011.

Martines, Lauro. *Lawyers and Statecraft in Renaissance Florence*. Princeton University Press, 2016.

Masini, Eliseo. *Sacro arsenale, ouero Prattica dell'Officio della S. Inquisitione, Ampliata*. Genoa, 1625.

Mayer, Thomas. *The Roman Inquisition: A Papal Bureaucracy and Its Laws in the Age of Galileo*. University of Pennsylvania Press, 2013.

Mayer, Thomas. *The Roman Inquisition: Trying Galileo*. University of Pennsylvania Press, 2015.

Mayhew, Robert. "Mapping Science's Imagined Community: Geography as a Republic of Letters, 1600–1800." *British Journal for the History of Science* 38, no. 1 (2005): 73–92.

McCabe, Richard A. *"Ungainefull Arte": Poetry, Patronage, and Print in the Early Modern Era*. Oxford University Press, 2016.

McIlvenna, Una. *Scandal and Reputation at the Court of Catherine de Medici*. Routledge, 2016.

McMullin, Ernan. "Galileo on Science and Scripture." In *The Cambridge Companion to Galileo*, edited by Peter Machamer. Cambridge University Press, 1998.

Migliorino, Francesco. *Fama e infamia: Problemi della società medievale nel pensiero giuridico nei secoli XII e XIII*. Editrice Giannotta, 1985.

Miller, Peter. *Peiresc's Europe: Learning and Virtue in the Seventeenth Century*. Yale University Press, 2000.

Millner, Stephen J. "The Florentine Piazza della Signoria as Practiced Place." In *Renaissance Florence: A Social History*, edited by John T. Paoletti and Roger Crum. Cambridge University Press, 2006.

Muir, Edward. *The Culture Wars of the Late Renaissance: Skeptics, Libertines, and Opera*. Harvard University Press, 2007.

Murphy, Kathryn, and Anita Traninger. "Introduction: Instances of Impartiality.'" In *The Emergence of Impartiality*, edited Kathryn Murphy and Anita Traninger. Brill, 2013.

Nature. "How Can Scientists Make the Most of the Public's Trust in Them?" (editorial). 1 February 2024.

Naudé, Gabriel. *Epigrammata in virorum literatorum imagines quas illustrissimus eques Cassianus a Puteo sua in biblioteca dedicavit cum apendicula variorum carminum*. Rome, 1641.

Norbrook, David. "Women, the Republic of Letters, and the Public Sphere in the Mid-Seventeenth Century." *Criticism* 46, no. 2 (Spring 2004): 223–40.

Norman, Corrie E. "The Social History of Preaching: Italy." In *Preachers and People in the Reformations and Early Modern Period*, edited by Larissa Taylor. Brill, 2001.

Offer, Avner. "Between the Gift and the Market: The Economy of Regard." *Economic History Review* 50, no. 3 (1997): 450–76.

O'Neill, Charles Edwards, and José María Domínguez, eds. *Diccionario histórico de la Compañia de Jesús*. Vol. 2. Institutum Historicum, SI, and Universidad Pontificia Comillas, 2001.

Ostrow, Stephen F. "Cigoli's Immacolata and Galileo's Moon: Astronomy and the Virgin in Early Seicento Rome." *Art Bulletin* 78, no. 2 (June 1996): 218–35.

Ovid. *Metamorphoses*. Translated by Frank Justus Miller, revised by G. P. Goold. Harvard University Press, 1916.

Pasachoff, Jay M. "Simon Marius's *Mundus Iovialis*: 400th Anniversary in Galileo's Shadow." *Journal for the History of Astronomy* 46, no. 2 (2015): 218–34.

Paton, Bernadette. "'Una Città Fatticosa': Dominican Preaching and the Defence of the Republic in Late Medieval Siena." In *City and Countryside in Late Medieval and Renaissance Italy: Essays Presented to Philip Jones*, edited by Trevor Dean and Chris Wickham. Bloomsbury, 1990.

Pera, Marcello. "The Role and Value of Rhetoric and Science." In *Persuading Science: The Art of Scientific Rhetoric*, edited by Marcello Pera and William R. Shea. Science History Publications, 1991.

Peters, Edward. "Wounded Names: The Medieval Doctrine of Infamy." In *Law in Mediaeval Life and Thought*, edited by Edward B. King and Susan J. Ridyard. University of the South Press, 1990.

Pettegree, Andrew. *Brand Luther: 1517, Printing, and the Making of the Reformation*. Penguin Random House, 2015.

Pettegree, Andrew. *Reformation and the Culture of Persuasion*. Cambridge University Press, 2005.

Pollmann, Judith. *Memory in Early Modern Europe, 1500–1800*. Oxford University Press, 2017.

Pollmann, Judith, and Erika Kuijpers. "Introduction: On the Early Modernity of Modern Memory." In *Memory Before Modernity: Practices of Memory in Early Modern Europe*, edited by Erika Kuijpers, Judith Pollmann, Johannes Müller, and Jasper van der Steen. Brill, 2013.

Poole, Ross. "Enacting Oblivion." *International Journal of Politics, Culture, and Society* 22, no. 2 (2009): 149–57.

Preston, Claire. *The Poetics of Scientific Investigation in Seventeenth-Century England*. Oxford University Press, 2015.

Pumfrey, Stephen. "Harriot's Maps of the Moon: New Interpretations." *Notes and Records of the Royal Society of London* 63, no. 2 (2009): 163–68.

Raphael, Renée. "Galileo's *Two New Sciences* as a Model of Reading Practices." *Journal of the History of Ideas* 77, no. 4 (2016): 539–65.

Raphael, Renée. "Making Sense of Day 1 of the *Two New Sciences*: Galileo's Aristotelian Inspired Agenda and His Jesuit Readers." *Studies in History and Philosophy of Science* 42, no. 4 (2011): 479–91.

Raphael, Renée. "Printing Galileo's *Discorsi*: A Collaborative Affair." *Annals of Science* 69, no. 4 (2012): 483–513.

Raphael, Renée. *Reading Galileo: Scribal Technologies and the "Two New Sciences."* Johns Hopkins University Press, 2017.

Raphael, Renée. "Reading Galileo's *Discorsi* in the Early Modern University." *Renaissance Quarterly* 68, no. 2 (2015): 558–96.

Ray, Meredith. *Margherita Sarrocchi's Letters to Galileo: Astronomy, Astrology, and Poetics in Seventeenth-Century Italy*. Springer, 2016.

Reeves, Eileen. "Complete Inventions: The Mirror and the Telescope." In *The Origins of the Telescope*, edited by Albert Van Helden, Sven Dupré, Rob van Gent, and Huib Zuidervaart. KNAW Press, 2010.

Reeves, Eileen. *Evening News: Optics, Astronomy, and Journalism in Early Modern Europe*. University of Pennsylvania Press, 2014.

Reeves, Eileen. *Galileo's Glassworks: The Telescope and the Mirror*. Harvard University Press, 2008.

Reeves, Eileen. *Painting the Heavens: Art and Science in the Age of Galileo*. Princeton University Press, 1997.

Reeves, Eileen. "Speaking of Sunspots: Oral Culture in an Early Modern Scientific Exchange." *Configurations* 13, no. 2 (Spring 2005): 185–210.

Ricci-Riccardi, Antonio. *Galileo Galilei e Fra Tommaso Caccini: Il processo del Galilei nel 1616 e l'abiura segreta rivelata dalle carte Caccini*. Successori Le Monnier, 1902.

Ridolfi, Carlo. *Le Meraviglie dell'Arte overo le vite de gl'illustri pittori Veneti, e dello stato*. Vol. 2. Venice, 1648.

Rietbergen, Peter. *Power and Religion in Baroque Rome: Barberini Cultural Politics*. Brill, 2006.

Rizza, Cecilia. *Peiresc e l'Italia*. Giappichelli, 1965.

Robey, Tracey E. "Damnatio Memoriae: The Rebirth of Condemnation of Memory in Renaissance Florence." *Renaissance and Reformation* 36, no. 3 (2013): 5–32.

Roffeni, Giovanni Antonio. *Epistola apologetica contra caecam peregrinationem Cuiusdam furiosi Martini, cognomine Horkij editam adversus nuntium sidereum: De quattuor novis planetis Gallilei Gallilei olim in Patavino Gymnasio publici Mathematici*. Bologna, 1611.

Rospocher, Massimo, and Rosa Salzberg. "An Evanescent Public Sphere: Voices, Spaces, and Publics in Venice during the Italian Wars." In *Beyond the Public Sphere: Opinions, Publics, Spaces in Early Modern Europe*, edited by Massimo Rospocher. Il Mulino/Duncker & Humblot, 2012.

Rublack, Ulinka. *Dressing Up: Cultural Identity in Renaissance Europe*. Oxford University Press, 2010.

Salvi, Lorenzo [pseud. of Vincenzo Figliucci]. *Stanze sopra le stelle e macchie solari scoperte col nuovo Occhiale*. Rome, 1615.

Sarasohn, Lisa. "French Reaction to the Condemnation of Galileo, 1632–1642." *Catholic Historical Review* 74, no. 1 (1988): 34–54.

[Scheiner, Christoph.] *De Maculus Solaribus et Stellis circa Iovem Errantibus Accuratior Disquisitio*. Augsburg, 1612.

[Scheiner, Christoph.] *Tres Epistolae de Maculis Solaribus*. Augsburg, 1612.

Schotte, Margaret E. *Sailing School: Navigating Science and Skill, 1550–1800*. Johns Hopkins University Press, 2019.

Schulte van Kessel, Elisja. *Geest en Vlees in godsdienst en wetenschap*. Staatsuitgeverij, 1980.

Schulz, Ronny F. "Myths of the Inventor: Inventing Myths in the Literary Concept of the Artistic Ingenium in Germany and Italy (1500–1550)." In *Allusions and Reflections: Greek and Roman Mythology in Early Modern Europe*, edited by Elisabeth Waghäll Nivre et al. Cambridge Scholars, 2015.

Segre, Michael. *In the Wake of Galileo*. Rutgers University Press, 1991.

Segre, Michael. "The Never-Ending Galileo Story." In *The Cambridge Companion to Galileo*, edited by Peter Machamer. Cambridge University Press, 2006.

Segre, Michael. "Vite di scienziati, vite di artisti." In *Firenze milleseicentoquaranta: Arti, lettere, musica, scienza*, edited by Elena Fumagalli, Alessandro Nova, and Massimiliano Rossi. Marsilio Editori, 2010.

Segre, Michael. "Viviani's Life of Galileo." *Isis* 80, no. 2 (1989): 206–31.

Seife, Charles. *Hawking Hawking: The Selling of a Scientific Celebrity*. Basic Books, 2021.

Serjeantson, R. W. "Proof and Persuasion." In *The Cambridge History of Science*. Vol. 3, *Early Modern Science*, edited by Katherine Park and Lorraine Daston. Cambridge University Press, 2006.

Servius, Petrus. *Dissertatio Philologica de Odoribus*. Rome, 1641.

Shank, J. B. "After the Scientific Revolution: Thinking Globally About the Histories of the Modern Sciences." *Journal of Early Modern History* 21, no. 5 (October 2017): 377–93.

Shapin, Steven. *The Scientific Revolution*. University of Chicago Press, 1996.

Shapin, Steven. *A Social History of Truth: Civility and Science in Seventeenth-Century England*. University of Chicago Press, 1994.

Shapin, Steven, and Simon Schaffer. *Leviathan and the Air-Pump: Hobbes, Boyle, and the Experimental Life*. Princeton University Press, 1985.

Shapiro, Barbara. *A Culture of Fact: England, 1550–1720*. Cornell University Press, 2000.

Shea, William R. "The *Discorso intorno alle cose che stanno in su l'acqua o che in quella si muovono*." Unpublished essay, available at the library of the Museo Galileo. Accessed 2018. https://opac.museogalileo.it/imss/resource?uri=341781&found=1&l=it.

Shea, William R. "Galileo's Discourse on Floating Bodies: Archimedean and Aristotelian Elements." In *Actes du XIIe Congrès International d'Histoire des Sciences: Paris 1968*. Vol. 4. A. Blanchard, 1971.

Shea, William R., and Mariano Artigas. *Galileo in Rome: The Rise and Fall of a Troublesome Genius*. Oxford University Press, 2003.

Smail, Daniel Lord. *The Consumption of Justice: Emotions, Publicity, and Legal Culture in Marseille, 1264–1423*. Cornell University Press, 2003.

Smith, Helen, ed. *Renaissance Paratexts*. Cambridge University Press, 2011.

Snyder, Jon R. *Dissimulation and the Culture of Secrecy in Early Modern Europe*. University of California Press, 2010.

Stella, Aldo. "Galileo, il circolo culturale di Gian Vincenzo Pinelli e la 'Patavina libertas.'" In *Galileo e la cultura padovana: Convegno di studio promosso dall'Accademia Patavina di Science Lettere ed Arti nell'ambito delle celebrazioni galileiane dell'Università di Padova 13–15 febbraio 1992*, edited by Giovanni Santinello. CEDAM 1992.

Stewart, Frank Henderson. *Honor*. University of Chicago Press, 1994.

Stolzenberg, Daniel. "The Holy Office in the Republic of Letters: Roman Censorship, Dutch Atlases, and the European Information Order, circa 1660." *Isis* 110, no. 1 (2019): 1–23.

Terence. *The Woman of Andros. The Self-Tormentor. The Eunuch*. Edited and translated by John Barsby. Harvard University Press, 2001.

Testa, Simone. *Italian Academies and Their Networks, 1525–1700: From Local to Global*. Palgrave Macmillan, 2015.

Torricelli, Evangelista. *Lezioni accademiche di Evangelista Torricelli*. Florence, 1715.

Torrini, Maurizio. "'Che il mio nome non si estingua': La morte di Galileo e le sorti della scienza." In *Firenze milleseicentoquaranta: Arti, lettere, musica, scienza*, edited by Elena Fumagalli, Alessandro Nova, and Massimiliano Rossi. Marsilio, 2010.

Tosi, Alessandro. *Portraits of Men and Ideas: Images of Science in Italy from the Renaissance to the Nineteenth Century*. Plus, 2007.

Traninger, Anita. "Taking Sides and the Prehistory of Impartiality." In *The Emergence of Impartiality*, edited by Kathryn Murphy and Anita Traninger. Brill, 2014.

Vaccalluzzo, Nunzio. *Galileo nella poesia del suo secolo*. Remo Sandron, 1910.

Valleriani, Matteo. *Galileo Engineer*. Springer, 2010.

Van Dam, Harm-Jan. "Poems on the Threshold: Neo-Latin *carmina liminaria*." In *Acta Conventus Neo-Latini Monasteriensis: Proceedings of the Fifteenth International Congress of Neo-Latin Studies*, edited by Astrid Steiner-Weber and Karl A. E. Enenkel. Brill, 2015.

Van Helden, Albert. "Galileo and the Telescope." In *The Origins of the Telescope*, edited by Albert Van Helden, Sven Dupré, Rob van Gent, and Huib Zuidervaart. KNAW Press, 2010.

Van Helden, Albert. "The Invention of the Telescope." *Transactions of the American Philosophical Society* 67, no. 4 (1977): 1–67. Reprinted with a new introduction in *Transactions of the American Philosophical Society* 98, no. 4 (2008).

Van Helden, Albert. "The Telescope in the Seventeenth Century." *Isis* 65, no. 1 (1974): 38–58.

Van Helden, Albert. "Telescopes and Authority from Galileo to Cassini." *Osiris* 9, no. 1 (1994): 8–29.

Van Helden, Albert, Sven Dupré, Rob van Gent, and Huib Zuidervaart, eds. *The Origins of the Telescope*. KNAW Press, 2010.

Van Krieken, Robert. "Celebrity's Histories." In *Routledge Handbook of Celebrity Studies*, edited by Anthony Elliott. Routledge, 2018.

Van Krieken, Robert. *Celebrity Society*. Routledge, 2012.

Vasari, Giorgio. *The Lives of the Artists*. Translated and edited by Julia Conaway Bondanella and Peter Bondanella. Oxford University Press, 1991.

Visser, Arnoud. "Scholars in the Picture: The Representation of Intellectuals on Medals and Emblems." In *Transmigrations: Essays in Honour of Alison Adams and Stephen Rawles*, edited by Laurence Grove, Alison M. Saunders, and Luis Gomes. Glasgow Emblem Studies 2011.

Visser, Arnoud. *In de gloria: Literaire roem in de Renaissance*. Inaugural Lecture, Utrecht University, printed in The Hague, 2013.

Volpi Rosselli, Giuliana. "Il corpo studentesco, i collegi e le accademie." In *Storia dell'Università di Pisa*, vol.1, pt. 1. Pacini Editori, 1993.

Walker, Claire, and Heather Kerr, eds. *Fama and Her Sisters: Gossip and Rumour in Early Modern Europe*. Brepols, 2015.

Westfall, Richard S. *Essays on the Trial of Galileo*. Vatican Observatory Publications, 1981.

Westfall, Richard S. "Galileo and the Accademia dei Lincei." In *Novita celesti e crisi del sapere: Atti del Convegno internazionale di Studi galileiani*, edited by Paolo Galluzzi. Giunta Barbèra, 1984.

Westfall, Richard S. "Science and Patronage: Galileo and the Telescope." *Isis* 76, no. 1 (1985): 11–30.

Westman, Robert S. *The Copernican Question: Prognostication, Skepticism, and Celestial Order*. University of California Press, 2011.

Wickham, Chris. "*Fama* and the Law in Twelfth-Century Tuscany." In *Fama: The Politics of Talk and Reputation*, edited by Thelma Fenster and Daniel Lord Smail. Cornell University Press, 2003.

Wilding, Nick. *Galileo's Idol: Gianfrancesco Sagredo and the Politics of Knowledge.* University of Chicago Press, 2014.

Xenophon. *Memorabilia. Oeconomicus. Symposium. Apology.* Translated by E. C. Marchant and O. J. Todd, revised by Jeffrey Henderson. Harvard University Press, 2013.

Zemon-Davis, Natalie. "Beyond the Market: Books as Gifts in Sixteenth-Century France." *Transactions of the Royal Historical Society* 33 (1983): 69–88.

Zemon-Davis, Natalie. *The Gift in Sixteenth-Century France.* Oxford University Press, 2000.

Zuidervaart, Huib. "The 'True Inventor' of the Telescope: A Survey of 400 Years of Debate." In *The Origins of the Telescope*, edited by Albert Van Helden, Sven Dupré, Rob van Gent, and Huib Zuidervaart. KNAW Press, 2010.

INDEX

Note: Page numbers in *italics* indicate figures.